COMMERCE AND MASS CULTURE SERIES

Edited by Justin Wyatt

Active Radio: Pacifica's Brash Experiment
Jeff Land

PACIFICA'S BRASH EXPERIMENT

JEFF LAND

COMMERCE AND MASS CULTURE SERIES
UNIVERSITY OF MINNESOTA PRESS
Mi
Lo

The University of Minnesota Press gratefully acknowledges permission to reprint lines from "Great Chinese Dragon," by Lawrence Ferlinghetti, from *Starting from San Francisco;* copyright 1961 by Lawrence Ferlinghetti. Reprinted by permission of New Directions Publishing Corp.

Published by the University of Minnesota Press
111 Third Avenue South, Suite 290
Minneapolis, MN 55401-2520
http://www.upress.umn.edu

Library of Congress Cataloging-in-Publication Data

Land, Jeff.
     Active radio : Pacifica's brash experiment / Jeff Land.
          p.   cm. — (Commerce and mass culture ; 1)
     Includes bibliographical references and index.
     ISBN 0-8166-3156-5 (hc). — ISBN 0-8166-3157-3 (pb)
     1. Pacifica Radio.   2. Alternative radio broadcasting —
United States.   I. Title.   II. Title: Pacifica's brash experiment.
III. Series.
HE8697.75.U6L36   1999
384.54'06'573 — dc21   98-56538

This work is dedicated with love to my parents, Arthur and Minna Land.

Contents

Congressional hearings in 1926 found legislators debating federal regulation of radio. Of urgent concern was securing a stable means of financing the programming in the five-year-old industry. Since 1921 income from the sale of radiolas themselves had subsidized the cost of the broadcasts. But the moment loomed when most households would own a receiver, eliminating the potential revenue from this source. The practice of selling airtime to advertisers to cover programming expenses had not yet spread to the majority of the nearly seven hundred radio stations in the United States. In fact, Secretary of Commerce Herbert Hoover, to whom the federal regulation of broadcasting had been delegated, was on record opposing radio commercials: "It is inconceivable that we should allow so great a possibility for service to be drowned in advertising chatter."[1]

Of the many criticisms of paid advertising in this period, one of the more interesting came in written testimony to the Congress from W. G. Cowles, head of the Chicago-based Zenith Radio Corporation. In 1926, Cowles viewed advertising as a threat to national security. If stations were to relinquish control over their programs to any paying sponsor, "Bolshevist propaganda will have a better chance in this country than ever before.... All radical thinkers, whether in politics, religion, or anything else, will fill the air with their efforts to poison the minds. This situation is intolerable."[2]

Congress continued its work for a year, fashioning the Dill-White Radio Act of 1927. This scaffold on which all subsequent telecommunication policy in the United States would rest contained but a single reference to advertising: any program sponsored by a business must identify that enterprise by name, so that wily, unnamed advertisers could not "subliminally seduce" the audience, as a later id-iom would put it. Thus, at the dawn of the mass media in the United States, the secretary of commerce publicly opposed the broadcasting of commercials, certain influential station owners equated advertised sponsorship with Bolshevist infil-tration, and Congress believed that a law was needed to guarantee that advertisers identify themselves.

Given what we now know about the relative significance of Bolshevism and commercialism in the history of this country's mass media, the foregoing sketch might seem farcical. By 1930, businesses, advertising agencies, and station own-ers had united in their appreciation of the opportunity radio provided for enticing "the most numerous and attentive audience ever assembled . . . in the quiet and in-timate atmosphere of the home[,] reached by the most natural channel for the ex-change of human thought, namely the speaking voice."[3] The mass media in the United States have for the past seven decades operated under the imperative to capture the largest percentage of this "most numerous and attentive audience" for their sponsors. Broadcasters have shunned not only Bolshevism but most contro-versial or complicated topics that might jeopardize their audience share. Critics from the 1920s to the 1990s have observed that the world framed and broadcast by commercial media bears scant resemblance to any actual state of affairs. As the antihunger organization Bread for the World recently asserted, there are more U.S. reporters whose full-time job is covering the New York Yankees than there are reporters in the entire continent of Africa.[4]

Active Radio: Pacifica's Brash Experiment, however, accentuates the positive: it investigates the heroic story of the *listener*-sponsored Pacifica radio network that against many odds established a noncommercial chain of five stations — in Berkeley, Los Angeles, New York, Houston, and Washington, D.C. Using the air-waves in a uniquely utopian manner, these stations have served as a voice pro-moting social justice, international solidarity, personal transformation, and cre-ative expression for five decades. Through its engagement with the culture and politics of the postwar world, the network has invited alliances with nearly every transformatory movement of the past fifty years, from the Beats and hipsters to the Weather Underground, from Salvadoran guerrillas to militant vegans.

By situating itself not as a neutral observer but as a committed participant within and across these movements, Pacifica hearkened to the demand of John Dewey that "the struggle for democracy has to be maintained on as many fronts as culture has aspects: political, economic, international, educational, scientific, artistic, and religious."[5] In its zeal, Pacifica has risked the loss of its licenses, had its transmitters bombed, seen its personnel arrested and jailed, and made errors of judgment and taste. Yet its tumultuous path has significantly increased the mass media's sphere of debate and performance.

Pacifica's programming contains a store of riches of the politics and culture of the past five decades. The network's tape archives and program guides only hint at the hundreds of thousands of hours of broadcasts on contentious political issues, analysis of emerging social movements worldwide, and live performances of contemporary musical, poetic, and dramatic compositions the network has presented. It has pioneered a number of media innovations: listener sponsorship, the use of the FM band, call-in radio, "underground" radio of the sixties, and "community" broadcasting in the seventies. In each of these experiments, Pacifica challenged the parameters of tolerable expression on the airwaves, living up to one of its original names: "Free Speech Radio." *Active Radio: Pacifica's Brash Experiment* teaches us that there are creative, practical alternatives to the commercial media and their hegemonic function.

This book makes extensive use of Pacifica's tape archives, but some important caveats should be mentioned. Although there are thousands of hours of tapes in the archives surveying the history of "nonviolence" and the myriad of "antiwar" movements since 1960, there are almost none from the earliest days when the founders expressed their own beliefs. For this reason, a range of ancillary sources — interviews, later retrospective summaries, and other writings — supplements the meager tape resources for one of my central concerns, namely, the manner in which Lewis Hill, Pacifica's founder, and his colleagues originally understood the relationship of a radio station to the pacifist movement of which they were a part in the 1940s.

A second lacuna is the brevity of my discussions of the music, drama, and poetry that from the origins to this day remain the greatest part of the network's schedule. A radical humanist and poet, Hill insisted that cultural programs would highlight Pacifica's schedule. The initial prospectus defined the goals of KPFA as "encourag[ing] and provid[ing] outlets for the creative skills and energies of the community"[6] through making the station a site of live performances of music and drama. It is not inconsequential that this goal was listed first among the varied political and cultural objectives that Hill and the others hoped to achieve, *before* one calling for international peace. Hill and the leadership of Pacifica never believed that lectures, news analysis, and discussion alone could bring about the transformation in consciousness pacifism demanded.

There are many people to acknowledge with heartfelt thanks. I wish to thank University of Oregon professors Carl Bybee, Julia Lesage, and Howard Brick for committing themselves to this project when it was a dissertation. Timely suggestions and gentle prods from all of them paved the way for its completion. My adviser and friend, Professor Janet Wasko, provided encouragement and expertise when most needed, working with me during her sabbatical. To begin research and writing on a topic such as this, I was indeed fortunate to have mentors who are no strangers to creative dissent and robust public debate.

Many former and present Pacifica broadcasters were unstinting in sharing with me their sense of the network and its history. Larry Josephson in particular was extremely kind. He made his extensive tape and documentary archive on Pacifica available to me, along with unlimited use of his Xerox machine. Special thanks are due to those in the Los Angeles tape archives who helped me get started with the research, sifting through the earthquake rubble in 1992 in order to find recordings of some vital programs buried under several feet of reels. Ed Morales and Adam Diamond provided important research assistance.

Cheyney Ryan, Dan Pope, Irene Diamond, and George Gessert read different chapters at various stages of completion, offering critical assistance along the way. Heartfelt appreciation to all of them.

Maya Chiah Sarah Diamond Land was born three weeks after I entered graduate school. This book and she have developed in tandem. Her curiosity made the completion of this project possible, keeping me sane through her daily reminders that there are more important things in life than the academy or the media.

For the first time in the history of American broadcasting, an
opportunity is offered listeners to maintain a radio station
distinguished by its candor, the unique quality of its programs,
and its freedom from commercials.

— *KPFA FOLIO*, AUGUST 1951, READ ON THE PROGRAM
"KPFA'S SIXTEENTH BIRTHDAY"

Active Radio begins with the early history of broadcasting in the United States,
outlining the circumstances in which a small, powerful group of corporations
came to control the vast majority of "our" radio channels. How did commercial
stations succeed in convincing both the government and early listeners that they,
not the educational, religious, and civic broadcasters, best served "the public interest"? During the 1920s, pioneering noncommercial broadcasters faced immense difficulty keeping their bearings as the federal government via its newly
formed Federal Radio Commission (1927) transferred nearly all broadcasting licenses to commercial stations.

Between 1920 and 1934, a contest over control of the airwaves occurred, one
often occluded from research and textbooks in media history. Without the full
means of producing a "consensus" they would over time obtain, corporate broadcasters needed to muster all their resources to convince the American listener and
the U.S. government that their oligarchic control of the airwaves was inherently
democratic and based on public service.

One vital site of this struggle occurred over licensing. Consider that approximately one quarter of the broadcast licenses distributed by the secretary of commerce between 1920 and 1925 were for noncommercial stations; many university
channels in particular offered their audience an eclectic if erudite schedule that
provided for an all too brief moment a viable, partial alternative to the entertainment of the corporate media.[1] In his important work *Telecommunications, Mass
Media, and Democracy,* Robert McChesney has argued that the corporate control
of the airwaves was not decided until the defeat of the Hatfield-Wagner Amendment

to the Federal Communications Act in 1934.[2] By the time Pacifica was founded in the late 1940s, the legacy of early noncommercial broadcasters and their incisive critiques of corporate media was but a faint echo.

Chapter 2 provides an overview of twentieth-century pacifism, the political vision that gave the network its name and ideals. The oldest dissident movement available for study and moral guidance, the struggle for peace has had an illustrious career, spanning the millennia with its plea for dialogue, negotiation, and trust. That human disagreement and competition need not be physically concussive seems a truth obvious beyond utterance; that warfare brings in its wake "an unending, universal mourning-wail of women, parents, orphans,"[3] should, one imagines, have led humans to beat their swords into ploughshares millennia ago. Yet as we see in a century during which the carnage of battle has claimed *hundreds of millions* of victims, mostly civilian, the quest for peaceful human coexistence remains an elusive goal. One of the central aims of this book is to document and evaluate Pacifica's passionate search for a "moral equivalent of war."

Pacifica was forged during an era of epochal transformation in the means of battle — the atomic age with its nuclear weapons and "security" based on mutual assured destruction. Lewis Hill, the young conscientious objector who guided the formation of Pacifica in the years following World War II, well understood that human technological genius had moved the apocalypse from religious myth to scientific challenge and government policy. After working for a Washington, D.C., radio station in 1943, Hill lamented the media's conspiracy of silence, entertaining and distracting rather than educating the public during and after World War II. How, he and his pacifist comrades wondered, might radio be deployed toward transforming our inclination toward violence and aggressive posturing — toward ending what the Quakers called "war and the occasions of war"?

In answering this question, Hill and his comrades molded a radical critique of the emerging military-industrial complex and national security state, support for social justice and civil liberties, and an abiding personal taste for avant-garde culture into the basis for daily radio programming. Through the responsible use of broadcasting, the men and women who established the Pacifica Foundation in northern California in 1946 were certain that radio was an indispensable means to educate "people of goodwill" about the futility of war, and further that broadcasting could and must be used as a means to hasten the end of all social injustice.

Its original articles of incorporation declared Pacifica's mission to be:

> In Radio broadcasting operations to engage in any activity
> that shall contribute to a lasting understanding between na-

tions and between the individuals of all nations, races, creeds, and colors; to gather and disseminate information on the causes of conflict between any and all such groups; and through any and all means compatible with the purposes of this corporation, to promote the study of political and economic problems and causes of religious and philosophical antagonisms.[4]

This yearning to diffuse the material and ideological antagonisms leading to war was and remains the center of Pacifica's enterprise.

Chapter 3 details the early days of this country's first successful listener-sponsored broadcast outlet, Berkeley's KPFA. Hill determined that listener sponsorship akin to magazine subscriptions might transform the way the mass media operate. More than that, he was able to put this insight into practice. A station supported by audience subscriptions would be free from both network and advertising exigencies. This in turn might open the ether to controversial political issues and the creative imagination of individual programmers.

Hill's single-minded devotion inspired a small band of artists, educators, media professionals, and pacifists in the San Francisco area. They donated an untold amount of time and expertise in the years directly after World War II to pursue the dream of a radio station dedicated to broadcasting creative expression and dissent. From its first programs in April 1949, KPFA sounded like nothing else on the airwaves, with its range of political discussion at the height of the Cold War, its celebration of literary and musical innovation, and its refusal to adopt tightly scheduled formats. This experiment, blending elements of anarchism, artistry, and egotism, was purchased at immense institutional and personal costs, leading to strikes, bitter personal attacks, and ultimately Hill's suicide in 1957.

Chapter 4 charts the expansion of the network from one to three stations, as new affiliates in Los Angeles and New York joined KPFA at the end of the 1950s. Pacifica's growth allowed it to further its goals of opening the airwaves to "promote the full distribution of public information; to obtain access to sources of news not commonly brought together in the same medium; and to employ such varied sources in the public presentation of accurate, objective, and comprehensive news on all matters vitally affecting the community."[5] Much of the network's programming lived up to these elevated goals; *Time* magazine ran a feature on KPFA entitled "Highbrow's Delight." At the same time, the network's dissident political positions and uncensored cultural offerings during this period led to relentless attacks by varied government forces. One particularly incendiary broadcast in 1962 was WBAI's public exposé of the FBI's illegal internal surveillance program, the first time the agency's spying activities had been publicly revealed. This interview with a former agent elicited immediate response from the bureau, which threatened key personnel at WBAI.

This pressure (repression, in fact) set in process a chain of events in which key personnel responsible for controversial programming were fired by the national board of directors, fearing that the FCC would follow the FBI's dictates and remove Pacifica's licenses. Some board members went so far as to consider complying with the FCC's request for loyalty oaths, an absolute anathema to almost all of the staff. These internal struggles, epitomized by the ongoing controversy over KPFA's legendary public affairs chief, Elsa Knight Thompson, in turn led to a series of strikes and work stoppages and continuing institutional traumas throughout the early 1960s.

Chapter 5 steps outside the chronological narrative to assess what lessons Pacifica teaches about the First Amendment. This chapter casts the network within the paradigm recently elaborated by legal scholar Steven Schiffrin, who has argued that

> the First Amendment has enlivened, encouraged, and sponsored the rebellious instincts within us all. It affords a positive boost to the dissenters and rebels. It has helped to shape the kind of people we are, and it influences hopes about the kind of people we would like to be.[6]

These sentences elegantly capture the particular Anglo-American libertarian tradition within which Pacifica situates itself, one that champions dissent ("rebellious instincts") as the lifeblood of democracy. That an enterprise devoted to peace should base its practices on free speech is hardly a historical anomaly — the American Civil Liberties Union began as a pacifist organization in World War I, for instance. Modern pacifism as well as modern rights-centered liberalism both find central ideological roots in the English Revolution. That lengthy seventeenth-century "event" proclaimed the sovereignty of personal conscience, from which sprang the proselytizing pacifism of the Quakers and the libertarian ethos of the Bill of Rights. This synthesis of religious and expressive freedom is codified in the language of the First Amendment.

Pacifica was born at a moment when official government secrecy in the name of "national security" was wreaking havoc on society as a whole and on the media directly. In this context, KPFA's programmers insisted that for a democracy, *no state policy is beyond media scrutiny*. Inspired by legal theorist Alexander Meiklejohn, the earliest broadcasts insisted that robust public discussion provides the foundation of democracy, educating citizens about national and global affairs.

Consider Meiklejohn's oft-cited description of the mass media in 1948, which reflects the bitterness of one who deeply feels the loss of radio's democratic potential:

> When [radio] became available, there opened up before us the possibility that, as a people living a common life under a common agreement, we might communicate with one another freely with regard to the values, the opportunities, the difficulties, the joys and sorrows, the hopes and fears, the plans and purposes, of that common life. It seemed possible that, amid all our difference, we might become a community of mutual understanding and of shared interest. It was that hope which justified our making radio "free," giving it First Amendment protection. . . .
>
> But never was a human hope more bitterly disappointed. The radio as it now operates among us is not free. Nor is it entitled to the protection of the First Amendment. It is not engaged in the task of enlarging and enriching human communication. It is engaged in making money. And the First Amendment does not intend to guarantee men freedom to say what some private interest pays them to say for its own advantage.[7]

Pacifica set for itself the goal to exemplify what a truly "free" radio does. (Pacifica's accomplishment in this realm reminds one, should that be necessary, of the unconscionable reticence of the commercial media — and subsequently PBS, for that matter.) Berkeley's free speech movement, one of the catalyzing moments of the sixties, has significant roots in the First Amendment vision KPFA had been broadcasting daily for nearly sixteen years in the Bay Area in studios several blocks from the campus. This great mobilization of students provides an important lens to understand the politics of Pacifica's early history.

Chapter 5 closes with a review of the most publicized event in the network's history: the FCC's censuring of a broadcast of George Carlin's "Seven Words You Can Never Say on Television" (sometimes called "Seven Dirty Words") monologue and the subsequent Supreme Court trial, which remains a landmark of First Amendment jurisprudence. This discussion moves the locus of our discussion of the First Amendment from the political to the pornographic, as it were. The dynamic practices of free speech on Pacifica challenge us to question not only the media's complicity in promoting the xenophobia leading to war but also the manner in which the invocation of "civility" in the cultural realm constricts creative freedom.

Chapter 6 picks up the thread of the sixties in light of the more theoretical argument of the preceding chapter. It highlights Pacifica's role in the origins of "community radio," focusing on New York's WBAI from the height of the antiwar movement in the sixties through its subsequent transformations from 1967 to 1977. During the 1960s, the network's libertarian stance evolved symbiotically with the movement to end the war in Vietnam and the growth of the counterculture. Through the decade, the programming at all Pacifica outlets evolved toward a greater agitation, marked by call-in programs, live broadcasts of rallies and demonstrations, and broadcast teach-ins.

WBAI, with an estimated weekly audience of more than a half million listeners in the late sixties, was the most successful outlet in Pacifica's history. At the cusp of the seventies, Third World, feminist, gay and lesbian, ecological, and new social movements of every persuasion clamored for increased access to WBAI's well-respected, widely heard microphone, all in the name of serving specific communities of listeners. In Hill's original vision, democratic broadcasting meant that the announcer should have complete autonomy over the content of the show; nonetheless, the strong identification of programs with highly particular groups (lesbian feminists, Gray Panthers) seemed to longtime listeners at odds with the definition of the audience as the "educated minority" of any given region. With this transformation from "free speech" radio into "community" radio, the politics of WBAI's programming shifted.

Pacifica in the early seventies, like much of the alternative and underground media, found itself struggling with a facile identification with the revolutionary idealism of the New Left and the exuberant hedonism of the Age of Aquarius. Different affinity groups, sects, and tendencies emerged, all seeing in Pacifica a novel means of reaching out to members. These dynamic, often radical new social movements included many groups vital today — feminists, environmentalists, gay and lesbian activists all coalesced in the early seventies around WBAI's microphone. The fierce commitment they brought to their politics flowed into struggles over programming and airtime. The difficulties in locating common ground across the range of emerging social movements led first to a work stoppage at KPFA in Berkeley in 1973 in controversy over the establishment of a "Third World" department. In 1977, at WBAI, various factions took over the station transmitter in the Empire State Building after a new program director attempted to revise the daily schedule.

Chapter 7 draws together some of the lessons from the struggles at WBAI. The democratization of Pacifica's programming and institutional arrangements was a challenging process, perhaps never fully accomplished. Studying the convulsions at WBAI serves as a useful chance to consider the opportunities and difficulties of developing political affiliation in the postmodern context. The struggles there also provide a way of dating the closure of the sixties, a complex "moment" during which contradictory tendencies of both liberation and constraint impacted a range of cultural and political activities. Pacifica's history in this period well bears this out.

The contests over access to the microphone at different Pacifica stations also usefully open questions of the nature and flexibility of the "public sphere," a unique space that became the site of much theoretical investigation more than thirty years ago. Pacifica's overall enterprise, and most interestingly where different ideals clashed, keenly delineates challenges that a democratic public sphere might face in the future.

The book closes by reconsidering the network's commitment to a world without war.

Active Radio's focus is Pacifica's first three decades, concentrating primarily on its stations in Berkeley and New York. It is during this period that the network's innovations in sponsorship, programming, and engaged social praxis forged a new form of broadcasting.[8] Within its historical narrative, *Active Radio* strives less for an exhaustive treatment of all the turmoil and triumphs of these years than for a distillation of the significance of the network, concentrating on the threefold themes of pacifism, free speech, and community. The unique synergy these ideals exhibited in the course of Pacifica's history serves to define a model of democratic communication. The peaceful, libertarian, and just society that has served as the utopian horizon for several generations of programmers was dialectically joined to the project of opening the mass media for a radically different, noncommercial form of broadcasting. There remains a strong yearning for diverse, uncensored, radical media. As a contemporary movement for "democratic communication" emerges to contest the extraordinary concentration and centralization of the media in the 1990s, my hope is that the early history of Pacifica will give some background and focus to this effort.

Pacifica is situated directly in the lineage of U.S. radicalism that has worked, relentlessly, to keep our founding revolutionary ideals of equality, liberty, and community alive. This democratic radicalism, running from Thomas Paine, through Walt Whitman, to John Dewey, Martin Luther King, and Saul Alinsky, has multiple inflections. (Hill in particular felt great personal affinity with Jefferson, Thoreau, and Dwight MacDonald.) A succinct, if conventional, reading understands democracy as the expansion of individual choice and opportunity in all areas, personal and political. In this light, one of the basic definitions of democratic broadcasting is the imperative of a great variety of voices, a wider opportunity for diverse opinion on the airwaves. Pacifica has pioneered this approach for the media in this country in the past fifty years.

Nonetheless, there are two ways this definition must be expanded to appreciate the fuller dimensions of Pacifica's efforts. The first concerns the way in which the radical tradition in the United States combines a unique blend of idealism and pragmatism impelled by the claims of personal conscience. Chapters 2 and 3 will return to this topic using the specific instance of the postwar nonviolent movement as an example of conscientious political practice based on individual response to social evil.

A second issue, noted by both critics and partisans alike, is that democracy in the United States is a uniquely unstable condition or "situation." Individual personality and social institutions undergo frequent transformation, either through the infusion of immigrants or through other forms of social ferment. First theorized by Tocqueville as the dialectic between liberty and equality, or in other categories such as the struggle between justice and excellence, or between community and autonomy, our political culture pulses with divergent, at times contradictory components. As Supreme Court Justice Oliver Wendell Holmes claimed in one of his most ardent defenses of the First Amendment, the bedrock of our political culture is neither free elections nor civil liberties per se. It lies, rather, in the necessary, perhaps thorough, disruption of convention encouraged by vigorous public dissent. "That, at any rate, is the theory of our Constitution. It is an experiment, *as all life is an experiment.*"[9]

John Dewey, whose philosophy was an early and important influence on Holmes's jurisprudence, wrote that democracy beckons toward the "continuous readjustment [of intellect] through meeting new situations . . . and the liberation of greater diversity of personal capacities."[10] Lifelong education ("growth") served as the key to liberation in Dewey's terms, keeping the intellect open and curious ("plastic"), inclined toward the "experimental" disposition that self-government invites, perhaps demands.

Dewey's voluminous political and educational theory (spanning nearly seven decades) provided the core vocabulary for the initial discussion of democracy and the mass media in this country and remains vital to the present. His works inspired thousands of educators and social critics in the 1920s and 1930s to consider what specific contributions the "miracle" of radio broadcasting could make in fostering the suppleness of intellect at the heart of a democratic culture. Following Dewey, these early broadcast reformers invoked a fairly capacious set of criteria for democratic mass media: radio could enthrall its audience with broadcasts that are "broad, wide, varied, and rich . . . enhanc[ing] open mindedness, and increased flexibility of thought and action."[11] At last a form of communication had emerged to serve a pedagogical, pragmatic vision of democracy. Opera, seminars, sermons, poetry, and political oration, all freely delivered directly to the living room, could serve to build a national community joined by intelligence and interest.

If there is a single paradigm linking the potential of broadcasting with the ideals of democracy, it is framed by the common set of terms: "flexibility," "plasticity," or "open-mindedness" — the goal of education in the broadest sense. The democratic citizen need not accept any "given"; we hold everything up for experiment and adjustment. In the words of a later writer, "it may be that the grandeur of democracy lies as much in its facilitation of aspirations and experiences as in anything else that may be true of it."[12] Democratic culture demands, and democratic media, through its varied, challenging programming, could provide this continuous readjustment of our horizons and aspirations.

(Alas, for these idealists struggling through the emergence of corporate media, the glorious civic and cultural possibilities that radio made available were smothered at birth, lost in the "advertising chatter": commercial radio entertainment was the antithesis of their hopes, serving the pecuniary interests of the few while inculcating in its audience a narrow, debilitating range of mindless habits — of consumption and passive reception. See chapter 1.)

Dewey proposed that the visionary who best articulated the meaning of democracy was Walt Whitman. Whitman insisted that democratic government must enhance individual creativity and the vibrancy of daily experience. For us to become the new nation that our founders imagined, Whitman demanded not only the liberation of slaves and women, the passionate love of comrades for each other, and fundamental economic redistribution, but most important a poet who could transform these "merely" social ideals into a celebration of the creative self. With its epic national poet heralding the way (and who better than Whitman to fulfill this role?), an "American personality" would finally emerge, freed from feudalism, "sloughing off surfaces, and from its own interior and vital principles, reconstructing democratizing society."[13]

Pacifica's daily blend of poetry and political debate often seemed guided by the argument of *Democratic Vistas* and the ecstasies of *Leaves of Grass.* Whitman's delirious, utopian challenge is one that Pacifica took to heart; this perspective frames *Active Radio*'s narrative. In more concrete political terms, Pacifica reminds us that hegemony is never fully secured; democratic culture's "experimental" disposition to received wisdom, habits, institutions, and traditions retains traces, if faint, of its radical birth. For the past five decades, Pacifica has defied odds and found a unique niche in this country's mass media, broadcasting sounds and ideas heard nowhere else. In doing so, the network has merged the outrage and hope that Whitman experienced in his survey of our "vistas" more than a century ago. Pacifica's programs have consistently confronted the gaps between the potential grandeur of intellect, spirit, and affiliation that democracy promises and the attenuated reality of postwar late-capitalist consumer society. Like its forebears in the American radical tradition, the network has not despaired in the face of such a gap but taken it as a challenge and the place to begin its work.

1. The Rise of Corporate Broadcasting

> Business succeeds rather better than the state in imposing its
> restraints upon individuals, because its imperatives are
> disguised as choices.
>
> —Walter Hamilton, quoted in James Rorty,
> Order on the Air

As the radiola craze swept the nation in the roaring twenties, corporate and educational broadcasters struggled to chart the destiny of the new medium. Both groups looked to Congress to regulate the distribution of licenses and keep some order in the chaos of rapid expansion. Between 1927 and 1934, the government and the emerging "mass media" industry, led by the newly formed National Broadcasting Company (formed in 1926), jointly worked to establish the principle that the nation's commercial stations best served the "public interest." With the federal government's redistribution of more than one hundred licenses from church, university, and civic stations to commercial stations during this period, noncommercial broadcasting in the United States was effectively smothered in the cradle.

This was neither a simple nor an uncontroversial process. With the creation of the Federal Radio Commission in 1927, the government began the delicate process of balancing the rhetoric that the airwaves were a public resource with the reality that the vast majority of "our" radio channels were actively placed in the hands of a small oligarchy of powerful private interests, broadcast corporations that continue to shape the media (and hence social and political) environment in which we live seventy years later. Understanding the early history of radio provides one context for measuring the extent of Pacifica's later accomplishment.

"This Magic Called Radio"

At the behest of the navy, Congress granted the control of wireless telegraphy to the secretary of commerce in 1912. Government oversight prevented the hundreds of amateur ham radio operators from crowding the airwaves with extraneous dots and dashes, a practice the navy claimed interfered with the use of "radiotelegraphy" in choreographing naval exercises. Acting under the penumbra of the Constitution's Commerce Clause, President Taft signed into law a Radio Act in 1912 mandating that whoever wished to transmit "radiograms" must first apply to the secretary of commerce. The secretary of commerce was authorized to grant a transmitting license to all those requesting one. Shipping companies as well as individual private operators applied for licenses, along with dozens of schools and universities where experiments in radiotelegraphy had been part of the physics curriculum for more than a decade. In all, more than eight thousand permits for sending wireless telegraphic signals were issued in the four years before the United States entered World War I. (Of course, the number of receivers was far higher than transmitters, by some estimates over one hundred thousand.)[1]

From the outset, both commercial and noncommercial operators played a central role in the development of what would later come to be called "radio." "Noncommercial" here refers to both educational institutions and the precocious amateur ham operators. In the first decades of this century, thousands of amateur aficionados (called "DXers" in the lingo) built their own crystal sets that could both transmit and receive wireless telegraphic signals. In 1910 the Wireless Association of America claimed ten thousand members; in 1912 the *New York Times* estimated that 122 wireless clubs held "over-the-air" meetings in Morse code on prearranged frequencies.[2] Their experiments hastened the development of the basic technological infrastructure that would flower into broadcasting within a decade.

World War I brought a blackout to this activity, shutting down all nonmilitary radio use of radiotelegraphy. Immediately after the war, thousands of amateurs, released from wartime restrictions, renewed their licenses and began experimenting with a variety of novel approaches to the medium. Although some had experimented with broadcasting both voice and music previously, the quality and range had generally been limited. Naval engineers, spurred by the war effort, had dramatically enhanced the capacities of transmitters, amplifiers, and receivers. In 1918, ham operators were thrilled by the extraordinary new opportunities for transmitting and receiving voice messages and music.

Although the war's end reopened the ether to the DXers, their day had passed. The war had proven the immense strategic value of wireless communication. Lessons learned from naval tactics now enabled shipping companies and distributors to coordinate their schedules far more efficiently using wireless. The navy, in alliance with different mercantile concerns, once again impelled the government to act, this time to hasten the establishment of a national radio corporation technologically sophisticated enough to contest the control of the British-based Marconi system in global wireless telegraphy. To protect national interests, many business, military, and government leaders worked to produce the Radio Corporation of America (RCA) to control all transatlantic and other international commercial telegraphic service from the United States. To achieve this, the secretary of the navy and other government officials gently twisted the arms of various patent holders in the radio industry to pool their trade secrets to expedite development of transmission devices, vacuum tubes, and receivers. In return for this patent sharing, different corporations were granted near monopoly control in their sector of the business.

It was in this milieu of postwar chauvinism, extensive amateur transmissions, and government-nurtured corporate mergers that the American mass media were born. Daily radio broadcasting was possible only because the vast grassroots network of operators had over time built hundreds of thousands of personal receivers. Regularly scheduled programming began, according to standard histories, in 1920. The first broadcasts served as a promotional supplement for a range of commercial, civic, religious, and educational enterprises. Hardware and department stores were the first enterprises to produce regularly scheduled shows aimed at convincing the uninitiated to purchase the electronic parts to construct their own radiolas. Radio's origins in the realm of telegraphy and "point-to-point" communication were indicated in the first name given the new phenomenon: the "wireless."

In 1920 and early 1921, department stores, newspapers, vacuum tube and radio set manufacturers, churches, and schools all requested licenses from the secretary of commerce to ply the ether with programming. Broadcasts of music, news, sports events, sermons, lectures, weather reports, community calendars, and comedy served to promote their goods and services. From fifty licensed wireless stations in the United States at the end of 1921, the field exploded to more than five hundred within a year. More than two hundred stations were run by radio and electrical manufacturers as a means of enticing customers to purchase their products. Schools, churches, and municipalities—that is, noncommercial enterprises—operated approximately 20 percent of these first stations.

Recall that RCA was initially formed to compete in the transatlantic telegraphic business coordinating shipping schedules and business messages. A statement from the corporation's 1922 annual report indicates how unexpected this paroxysm of broadcast (or "wireless telephony") activity was:

> At the time your corporation was formed in 1919, for the purpose of building up a world-wide international commu-

nication system, wireless telephony had not passed out of the experimental state, and it was not at that time foreseen that the broadcasting art would ever reach the high point of popularity that it has in the past year. The engineers and scientists had anticipated the development of wireless telephony for communication purposes, but *no one had visualized the phenomenal expansion of wireless telephony as used today for broadcasting.*[3]

Almost immediately, the "social destiny of radio" became the subject of much glorious anticipation. Broadcasting, following electricity, the telegraph, and the telephone in the previous century, was heralded as a technological miracle capable of immense beneficial social transformations. Traveling on etheric channels, invisible electronic waves radiating from the heavens provided a seemingly unlimited bounty of entertainment and education. Anyone could have the best seat in the Metropolitan Opera, for free!

According to RCA president General J. G. Harbord, in a typical panegyric, radio was a boon to democracy, freeing the citizen from the "contagion of the crowd." The solitary voter, listening to politicians in the privacy of his or her own living room, need not be a slave to mob enthusiasm but now was free to make political judgments based "solely to the logic of the issue."[4] Preachers would convey the divine message to those who refused to attend church. Instantaneous international communication would end war. For all these rosy predictions, soon contradictions would emerge between the humane and desirable ends toward which radio communication might be used and the business imperatives of the corporations who held controlling interest in the means of communications.

At first consumers, piqued by the desire to participate in this highly publicized fad, sparked a huge boom in sales of radiola sets and components. Profits from the sale of hardware in turn sponsored a great deal of initial programming. But at the same time, the sheer novelty of reception itself would not sustain continued listening. From the start, the appeal of broadcasting lay in continuously new programming. Consequently, the issue of who would pay for the expense of wireless telephony productions was a question from the start. Whereas in almost every other country the government in some fashion directed and financed the emergence of radio broadcasting during this period (much as the state had overseen previous postal, telegraphic, and phone services), the United States, with its tradition of a free press and regulated but private ownership of point-to-point communication, proved to be different.

Listeners, government officials, and radio operators alike tended to agree that those with a direct stake in radio should develop programming for it. The American audience was habituated from the start to the notion that radio programs were free public benefits provided by businesses, newspapers, schools, churches, and other public institutions as a type of self promotion — with the state itself playing only a minimal oversight role in license distribution. In a significant semantic shift, profit-motivated corporations seemed to serve the "public" interest by providing free, easily obtainable programming. This merging of public and private interests served the broadcasting corporations admirably when more direct government regulation loomed. Broadcasting in the United States was thus not only born without any public funding but would also come to fuse a rhetoric of public service, the democratic free market, and later the Bill of Rights, to elude most forms of public oversight once licenses were secured; the "Public Service Broadcasting System" (David Sarnoff's original name for NBC)[5] would ultimately differ from any other system in the world in the privacy of its control.

Private or corporate control over the programming was not originally synonymous with advertised sponsorship. Indeed, for the better part of radio's first decade, even partisans of corporate ownership of licenses, such as secretary of commerce Herbert Hoover, were not convinced that paid commercials should be the financial base of broadcasting in the United States. This situation was transformed in large part owing to the fragility of the patent-sharing agreement that formed RCA.

American Telephone and Telegraph (AT&T) was part of the consortium that established RCA in 1919. The agreement provided AT&T a monopoly in marketing its technologically advanced AM transmitters to radio stations, but the company was blocked from producing or selling less expensive home radio components, the economic backbone of the industry for the first several years. Backed by vast capital resources, and not needing to promote its monopoly on telephone service, AT&T's entry into broadcasting itself was unique. Its New York station, WBAY (later called WEAF, and finally WNBC) was modeled as a "phone booth on the air." Unlike other outlets, the station would produce few programs but would rent its studio, one of the finest in the land, charging for time, just as AT&T charged for its phone service. Anyone could use the studio, with a "toll" placed on the time the users spent on the air. For $50, one could speak for ten minutes. The first paid (or commercial) program in U.S. broadcasting took place in August 1922 on WEAF: a ten-minute pitch for Hawthorne Court apartments in Jackson Heights, Queens, which made much reference to Hawthorne's love of the outdoors and tidy homes. After several months, with only three hours of studio time purchased for $550, WEAF considered sending its ad salesmen back to the phone company.[6]

However, AT&T, having invested a considerable amount in the equipment for its studio, persisted a while longer, arguing before the licensing authority (Secretary of Commerce Herbert Hoover) that its "public service" venture required a new channel of its own, a highly valuable "clear" channel that would be free of any interference "from the clamor of the self-serving voices" of the other stations.[7] In 1924 all broadcasting officially occurred on only two frequencies. Almost all stations were required to share time during the day at 360 cycles. In the first years of broadcasting, the low power and limited range of the transmitters prevented this from becoming an insuperable problem, but by the mid-twenties, interference from distant stations had led to operators (illegally) roaming up and down the bandwidth searching for better frequencies. AT&T's successful petition before Hoover for a "clear channel" at 400 cycles that could be used only by WEAF was a huge boon. Its signal would now be far clearer, and WEAF need not share its broadcast day with other stations.

WEAF realized that businesses that did not wish to produce their own shows might be willing to subsidize studio-based productions. Thus, the original phone booth model was somewhat modified: WEAF, like other stations, would produce its own programming; it would then work to sell the audience of these programs to advertisers. Tying sponsors to weekly in-studio programs began in 1923 when the "Browning King" Orchestra made its debut on WEAF. The only reference to the fact that it was a clothing retailer who underwrote the music was the name of the show itself. Simultaneously with its experiments in commercially sponsored entertainment, AT&T was establishing a system of linkups with its various stations around the country, sending New York–originated programming out via its "long [phone] lines." This use of the telephone infrastructure immeasurably amplified the potential size of the audience a sponsor could reach, and hence the value of the time that was sold. At the same time, it established the precedent for centrally controlled national programs, an anomaly in a broadcast universe dominated by regional and local productions aimed at a small, geographically limited audience. As the "toll" for access rose to more than five hundred dollars an evening hour, few private individuals could enter the "phone booth of the air." With the charging for airtime and the establishment of chains of stations using the same program simultaneously, AT&T developed the basis for the mass media we have today.[8]

The "American System," as this corporate-based, commercially sponsored national structure came to be called, was a hybrid "network" of individually licensed stations whose titular autonomy successfully shielded the industry for years from

the taint of monopoly. Stations within the network had access to programs produced and distributed by New York–based NBC, linked by special AT&T telephone wires, a service that cost NBC about one million dollars a year. From its
central location, NBC marketed time and ears to businesses seeking a national
audience while paying its affiliates to air both the commercials and the programs
they sponsored. During daytime hours that had no "sponsored" programs, stations
would pay the network for the use of "filler" material — "sustaining" programs.
This arrangement, which enabled local network affiliates to have access to a *full
daily schedule* and a steady source of revenue, would shortly have significant implications for federal licensing and frequency distribution policy.

As the trickle of broadcast advertisements grew into a flood, the public was
not altogether pleased. In the late twenties, many listeners wrote to Congress and
the networks to complain about the commercial onslaught. The most vehement condemnation of the rapid commercialization and centralization of programming came
from educators. They claimed that the control of radio had rapidly devolved onto a
potentate of media moguls, producing mindless entertainment solely to attract the
largest audiences for their sponsors. In response, the radio industry could claim
that licenses were "owned" by hundreds of independent stations. While critics argued that advertisers ultimately determined the nature of programming by demanding the largest possible audience, the government, citing its widespread distribution
of licenses, countered that it had taken effective steps to prevent monopolistic
practices in the industry, thereby ensuring the nation essential First Amendment
practices in the electronic media.

This vocal backlash to the commercialization of the ether demanded a stream
of public relations maneuvers from the broadcast corporations. Advertisements
should be seen not as enticements to mindless consumption but as a godsend that
saved radio from the autocratic hand of government. As put with exemplary hon-
esty by Merlin Aylesworth, the president of the four-year-old National Broadcasting Company, in a statement from 1930 worth carefully considering:

> It was a kind fate that caused commercial broadcasting to
> see the light of day in America, the new world, the land of
> opportunity, the haven of advertising and publicity. Having
> created a vast audience, and following in the footsteps of
> AT&T, the newly formed [NBC] naturally turned to spon-
> sorship as the solution of its economic existence. Instead of
> looking upon the growing audience as a liability, this grow-
> ing audience became a valuable asset. Here indeed was the
> most numerous and attentive audience ever assembled. It

could be reached in the quiet and intimate atmosphere of
the home. It could be reached by the most natural channel for
the exchange of human thought, namely the speaking voice.
And so the sponsored program received consideration.[9]

One does well to ponder the implications in defining the audience as a valuable "asset." Fortune of the pecuniary sort indeed smiled on those able to capitalize on this situation, a blend of fortuitous circumstances that Aylesworth could only attribute to "kind fate." Corporate energy and laissez-faire governmental policy had established the conditions for the mass production of a new and fabulously valuable product: "the listening audience." Conjured up by the "most natural channel for the exchange of human thought," this vast aggregation of people, potentially global in size, could be reached in the quiet and intimacy of their own homes. Later someone would remark that getting a radio license at this time was like receiving a license to print money.

THE FEDERAL RADIO COMMISSION

During the first ten years of broadcasting, a tripartite system slowly emerged: the public "owned" the airwaves; the government managed, or regulated, the channels at the behest of the people; and private broadcasters generally controlled the scheduling and program content. In 1925 there were approximately 450 licensed stations in the United States, of which 125 were noncommercial. (This latter figure is the highest percentage that nonprofit stations would obtain.)[10] When the District Court for the Northern District of Illinois ruled in *U.S. v. Zenith* in 1926 that Secretary of Commerce Hoover's role in licensing stations was unconstitutional, chaos ensued. More than one hundred new, unlicensed, stations almost immediately began broadcasting. For the moment, no person or agency was legally empowered to determine channel, power, or airtime. These new broadcasters saturated all available frequencies, making reception a random and difficult procedure. For a frantic year, the federal government and radio operators with large investments in their hardware struggled to reestablish order on the air.

By the mid-twenties, the federal regulation of radio was far more complex than the basic distribution of licenses for point-to-point telegraphy mandated by the 1912 Radio Act. Thirty-one different bills and resolutions had been submitted to both houses of Congress from 1921 to 1926 addressing various aspects of broadcast policy; all stalled at the committee level, owing in large part to the success that Hoover had in convincing Congress that his Commerce Department was the natural home for all federal regulation. However, the Zenith decision made it clear that a new mandate was needed.

The attempt to produce a revised legal rationale for federal regulation in 1926 following the Zenith decision occurred during a period when some of the more sordid details of the Teapot Dome scandal had recently come to light, leading to public outrage and the call for Congress to demonstrate far more judicious husbanding of all the public's limited resources — such as etheric radio channels. Between the demands for the regulation from the station owners seeking relief from the chaos of a truly free market, and the government's mandate to act with renewed vigor in policing national resources, a constellation of forces impelled the creation of the Federal Radio Commission.

The operating assumptions for the emerging consensus on radio regulation in light of the Zenith decision were stated clearly by Representative Wallace White, principal author of the 1927 Radio Act:

> We have reached the definite conclusion that the *right of all our people to enjoy* this means of communications can be preserved only by the repudiation of the idea underlying the 1912 law that anyone who will, may transmit, and by the assertion in its stead of the doctrine that the right of the public to service is superior to the right of any individual to use the ether.[11]

The Dill-White Radio Act of 1927, which established the Federal Radio Commission and its licensing regime, became the core of all subsequent broadcasting regulation. As White's quote intimates, federal regulation identifies "enjoyment" as the core experience of radio while positing that access to the ether must be limited to safeguard the fundamental ability to receive signals. Almost nothing in the emerging legislation addressed either advertising sponsorship or corporate control of the networks, the material basis of the American System of broadcasting.

The first consequence of the 1927 act was the suspension of all existing licenses and the requirement for every station's reapplication. In a time-honored tradition, the act mandated that Congress establish a committee to oversee this relicensing of the stations. Aware that one of the main legal complaints against Hoover's authority over licensing was the charge of "capriciousness," the Radio Act, in one of its most consequential gestures, enunciated the criteria of "public interest, convenience, and necessity" to guide the allocation of the soon to be scarce, valuable licenses.

The Radio Act's phrase "public interest, convenience, and necessity" had its immediate origins in natural resource legislation of the late nineteenth century. In this legal context, the individual *interest* in power, heat, or water could be transformed into state policy regulating utility rates in such a way as to meet the aggregate needs of an expanding population with relatively minimal controversy. It is important to distinguish between the bodily desire for warmth or light, which preexisted the construction of mines and aqueducts, and the "necessity" of clear radio channels. The "listening" public claiming a "right" to "enjoy" radio was fundamentally an artifact of the changing technology. It was far from a homogeneous group that could be identified by specific material characteristics such as desire for warmth.

In this context, Dewey's discussion of the public, and its problems, in 1926 is germane:

> The machine age has so enormously expanded, multiplied, intensified and complicated the scope of the indirect consequences, [has] formed such immense and consolidated unions in action, on an impersonal rather than a community basis, that the resultant public cannot identify and distinguish itself.[12]

The listening audience was a clear paradigm of Dewey's "immense" and "impersonal" unions. Ceding control of communication to corporations that would use "the most natural channel for the exchange of human thought" to profit in "this haven of advertising and publicity" would only exacerbate the problem of the erosion of norms on which to guide community activity. Commercially dominated media would have no interest in participating in the complex, experimental, and educative process of fostering intelligent, communal interaction: there would be no profit in such an endeavor.

Federally established broadcast standards were the practical results of the Radio Act's interpretation of the meaning of public interest. The most significant consideration in license distribution was the number of hours of guaranteed programming a station could offer. Noncommercial civic, educational, and church broadcasters found themselves at a severe technical or financial liability competing with the affiliates of the chains who, fed with unlimited programs from their parent studios in New York, could easily fill the day with the Happy Wonder Baker's Quartette.[13] Of central importance was the commission's "General Order 40" issued in 1928. This administrative ruling divided the ether into the AM band of today. Ninety channels would be used. Of these, the most important by far were the forty "clear channels" licensed to forty single stations. Six hundred other outlets would be required to share the remaining fifty channels. Affiliates of NBC and CBS received most of the clear channels. The percentage of total airtime of network affiliates and large independent urban commercial stations rose from 30 to 70 between 1928 and 1931.[14]

Within this context, the government also established the precedent that licenses could automatically be retained by stations that had already secured one; once obtained, broadcast licenses could essentially be held in perpetuity, leading to their enormous value. The statutory argument that the government holds ultimate jurisdiction over the licensing process has enabled the myth of public ownership of the airwaves to persist over the years even as it became abundantly clear that the licensees — station and network owners — have de facto control over the spectrum.

This redistribution of channels was a vital blow to noncommercial stations, grievously limiting the opportunity for the "public" to become "interested" in the educational, cultural, and religious programming these stations produced. E. Pendleton Herring, in a widely quoted article in the *Harvard Business Review* of 1935, wrote:

> The point seems clear that the FRC has interpreted the concept of public interest so as to favor in actual practice one particular group. While talking in terms of the public interest, convenience, and necessity, the [Federal Communications] commission actually chose to further the ends of commercial broadcasters. They form the substantive content of public interest as interpreted by the commission.[15]

An occasional politician or professor might be allowed access to the microphone as "decoration on a station's record when the time comes for license renewal ... [but] underlying all considerations is the necessity of eliminating any element that might lessen the usefulness of the station as a device for attracting the buying public."[16]

From $4 million in 1926, the sales of advertising rose to over $15 million in 1929 and climbed to almost $112 million in 1935. Over this period, individual programming lengths shifted to accommodate the needs of advertisers. Shorter musical and dramatic presentations came to dominate the sound of radio in the "variety show" format. Announcers found ever more opportunity to thank the generous support of the sponsor, and to read the "spot" ad that promoted products or services unrelated to the program that preceded or followed it.

The great financial windfall that advertising brought to the major commercial stations led to the consistent technical upgrading of radio studios. This in turn increased the production quality of the music and drama programs the chains could offer their affiliates. It is in this dialectic between enhanced production quality and variety, underwritten by burgeoning commercial sponsorship, that the ideology of popular entertainment as public service finds its grounding.

ALTERNATIVES

Educators had gravitated toward radio since the late 1890s. In 1906 Cornell University offered courses in radio and communication engineering. Later, and more important, radio broadcasting provided the host university with the ability to broadcast lectures and cultural programs, generating favorable publicity for "ivory tower" institutions. KOAC, broadcast from the Oregon State Agricultural College, like many of these early institutionally based stations, was extremely popular in the largely rural areas of the state in the twenties. There was a significant audience for the weather reports, football, lectures, agricultural information, household hints, and student orchestra performances that the college-based station produced.[17] This particular outlet wisely opened its facilities to the governor and other state officials, who in turn guaranteed KOAC funding when the struggle over licenses began in earnest in the late twenties.

As previously described, after 1927, federal definitions of the "public interest" conformed to the capacity of the most well funded outlets — stations able to upgrade their hardware yearly and pay for talent to fill the airtime. Noncommercial stations competed on a severely lopsided playing field when applying for licenses before the FRC. University or civic support would never equal the tens, hundreds, then thousands of millions of dollars of advertising revenue that commercial stations generated; by the end of the decade, educational broadcasters arguing for the necessity of retaining their licenses while broadcasting for only part of the day were cast as a "special interest."

The imperative to sustain daily programming at suitable levels of transmission took its toll, with station after station ceding their licenses after the passage of the 1927 Radio Act. The educators were furious but found themselves caught in desperate cycle. As Joy Morgan wrote in 1931, synopsizing the bind into which the FRC had trapped noncommercial broadcasting:

> The practice of squeezing these stations off the air ran something like this. First, they would be given less desirable fre-

quencies, the more desirable being assigned to commercial and monopoly groups. Second, they would be required to divide their time with some commercial interest. Third, they would be required to give a larger share of their time to commercial interests. Fourth, they would be required to meet some new regulation involving costly equipment—often a regulation essentially right in itself, but applied with such suddenness as not to allow time for adjustment in the educational budget. Fifth, the educational station would be required to spend, on trips to Washington for hearings before the Federal Radio Commission and lawyer's fees, the money which should have gone into the development of personnel and programs.[18]

If these machinations did not succeed, the threat of lengthy, expensive litigation was always a last resort in those rare instances in which the "commercial broadcasters have not been able to browbeat the federal regulatory body into squeezing the 'long-hairs' out of the picture."[19]

During this period when radio was taken over by commercial interests and its "natural" basis in advertising was promulgated as a "kind fate" from which all might benefit, there were alternative models. The most significant was the British Broadcasting Corporation (BBC) system, financed by taxes on receivers. Broadcasters in the United States spent a great deal of ideological labor attacking the BBC in the late twenties. The National Association of Broadcasters published a scathing report on the BBC that exaggerated the number of listener complaints, alleged censorship on the part of the British state, criticized the lack of variety of programs, and, most emphatically, asserted that people have a right to choose their entertainment freely, a right potentially ignored by the British government "dictatorship" over programming, with its stress on educational and civic values.[20]

Many congressmen and a good deal of the American public recognized that for the first decade of radio broadcasting, educational, religious, and civic outlets had provided important and interesting programming not available on the commercial channels. All noncommercial stations could not be broadly painted with the taint of being "propaganda stations." With the onset of the New Deal after the election of Roosevelt in 1932, and vast public antagonism toward "big business," many educators believed their time to regain the airwaves had arrived. Unfortunately, the effort to force Congress to secure a substantial number of channels for commercial-free stations failed in the face of the networks' massive lobbying and some tactical confusion by the educators. The central question concerned whether these stations could solicit any paid sponsorship at all if they were to be designated "noncommercial." The most well known of their proposals was the Hatfield-Wagner Amendment to the Communications Act of 1934, calling for the government to reserve 15 to 25 percent of the frequencies for noncommercial stations.

The main ideological argument put forward by the lobbyists for the commercial stations centered on the limited sense of public responsibility felt by noncommercial broadcasters. Noncommercial stations, free to broadcast whatever they chose and lacking certain (profit) motives, could potentially fall prey to special interests. This insidious logic convinced certain key congressmen to vote against the Hatfield-Wagner Amendment. It was one thing to acknowledge the propriety of *some* noncommercial use of the airwaves, on which almost everyone agreed, but how would a fixed percentage of the channels ever get divided? For example, if major religious organizations requested free airtime, then, worried one congressman, what of "the Hindus, other infidels, . . . and the national association of atheists. They perhaps would want some time."[21] (This difficulty would not be a humorous exaggeration in the later struggles at Pacifica. Which groups deserved access to the microphone at a community station that ostensibly guaranteed democratic access for all?)

The failure of the Hatfield-Wagner Amendment in the midst of the New Deal frenzy proved to be the swan song for the first generation of broadcast reformers. Noncommercial stations saw their call letters, frequency, and hours shifted continuously by the FRC. Unable to cultivate an audience, unsure what their future chances would be, most noncommercial stations simply petitioned the commission to withdraw their licenses, seeing that they had no future in broadcasting. By 1936, of the 618 stations in the United States, there were fewer than forty noncommercial outlets. Thus, under government stewardship, the educational possibilities of broadcasting were whittled down to size and eventually left in the hands of such agents of democracy and culture as Du Pont Chemicals, Philip Morris, and General Electric.

The example of the highly touted program in civics *You and Your Government* might well serve as a bellwether of the "public interest" considerations of the commercial networks. Initiated in 1932 at the behest of NBC in loose concert with the American Political Science Association, this program over its first four years enlisted the talents of teachers and politicians nationwide to provide provocative commentary on a number of topics from taxes to the Constitution. Originally guaranteed a permanent prime evening half hour of Tuesday at 8:00 P.M., *You and Your Government* was a sustaining program, which NBC produced and distributed to its affiliates with no commercial sponsorship in order for them to both fill their schedule and meet their public interest mandate. For several years, this half-hour program provided excellent public relations and developed a substantial audience. Much as Pacifica would claim fifteen years later, the program's promotional material proudly explained that it was geared for an educated, not a mass, audience.[22]

However, by 1935 almost all evening hours on the network had found a sponsor; one hour of network prime time was valued at $14,250. *You and Your Government* had been pruned to fifteen minutes and shunted to a 7:30 P.M. time slot, the first of several time shifts. By this time, the modern contract between network and affiliate had been instituted: local stations were bound to accept all sponsored programs but could pick and choose from the variety of sustaining programs, retaining a certain autonomy over the unsponsored time slots. Those stations that once proudly used *You and Your Government* as a means of touting their public service found it more financially prudent to use the valuable early-evening time for locally based commercial programming, mostly popular recorded music. Many stations claimed they wanted to broadcast the show but needed the extra income from their sale of the nonnetwork time; more and more exercised their option to cancel broadcasting this erudite program.

Even more disheartening, the producers were not given any warning which stations around the country would be broadcasting the program at any given time. This made it next to impossible to promote the show, part of whose prestige was based on making ancillary pamphlets and teaching materials available to the listeners and other educators. Finally, as with so many relationships gone sour, NBC simply stopped returning the producers' phone calls or answering their letters, allowing the show to wither away. As the producers of the program wrote in an obituary published in 1937:

> [NBC's] educational department is weak and dominated by the business department. The shifts in time to which the "You and Your Government" program was subjected were brought about by the sales department of the NBC, over the protests, however feeble, of their own educational director. The organization and personnel of the Company have changed considerably in the course of our four years' relations, and it appears that the ideal of a "well rounded program service to the American public" has gradually been submerged by the pressure of financial concerns. . . . All this illustrates very well the precarious position of education in general, and civic education in particular, on the air. One of the best developed,

most varied, and extensively promoted educational programs
was cut off chiefly on the point of its successful conduct for
four years. This is whimsical and fantastic and would be
amusing except for its serious consequences.[23]

In the years following the consolidation of media control by corporations, there
was intermittent discussion of the manner in which the licensees were fulfilling
their public responsibility. A survey of network daily programming from the
1930s gives a clear, if rueful, snapshot of the practical manner in which the radio
industry exercised its First Amendment freedoms. In 1937, for example, more
than four hundred shows are listed by title and airtime, of which fewer than sixty
can, even with great leeway, be considered other than pure entertainment.[24] The
warnings of early critics of commercial sponsorship—that "the stimuli of art,
science, religion are progressively expelled to the periphery of American life"[25]—
seemed well borne out.

This evisceration of complex, intelligent, and controversial programming within
the media was not a natural process. Restrictive federal regulations and massive
federal lobbying campaigns would work in tandem to fuse the broadcasters' particular concern for advertising revenue and audience share with the "public interest, convenience, and necessity." In the end, this alchemy was made possible by
the corporate control of the fabulous, novel product that the broadcasting industry produced: freely distributed popular entertainment.

2. Lew Hill's Passion and the Origins of Pacifica

I do not know what is true. . . . But in the midst of doubt, in the collapse of creeds, there is one thing I do not doubt, . . . that is that the faith is true and adorable which leads a soldier to throw away his life in obedience to a blindly accepted duty, in a cause which he little understands, in a plan of campaign of which he has no notion, under tactics of which he does not see the use.

—Oliver W. Holmes, "The Soldier's Faith," in
The Mind and Faith of Justice Holmes,
selected and edited by Max Lerner

The ideal of a world without war led to Lewis Hill's involvement in the movement for revolutionary nonviolence during and after World War II, and subsequently to his founding of Pacifica. Long associated with religious conviction and individual witness, pacifist ideology and strategy underwent a dramatic transformation in the twentieth century. The unfathomable carnage of World War I and the luminous example of Mahatma Gandhi combined to forge a more politicized and oppositional form of struggle—"radical pacifism"—based on principles of active nonviolent resistance to war and to the social circumstances that engendered violence. Hill and other founders of Pacifica were deeply involved in this movement; their collective experience as pacifists during World War II molded the vision they pursued for their radio station.

"The Moral Equivalent of War"

At the dawn of the twentieth century, the international peace movement flourished as never before. Individuals and organizations on both sides of the Atlantic agitated in favor of disarmament and for a World Court to arbitrate regional or national conflicts. In this context, those struggling against war sought a dynamic, "scientific" grounding for their political beliefs, finding the older religious vocabulary anachronistic. As the president of the League of International Peace and Liberty put it in 1901, first giving the peace advocates a name they continue to use:

> Our great party needs a name; we have no name and this
> deficiency impedes our progress considerably. We are not

> passive [i.e., religious] types; we are not only peace makers,
> we are not just pacifiers. We are all those but also something
> more — we are *pacifists* . . . and our ideology is *pacifism*.[1]

This desire to demonstrate that "pacifists" were "something more" than just "paci-
fiers" was made evident in the spectacle of grand peace conferences held in 1899
and 1907 at The Hague. These events addressed questions of trade in arms, the
rules of war, and the construction of a World Court. An international rule of law,
derived from the Enlightenment tradition exemplified by Kant's essay "Perpetual
Peace," served as the practical, "rational" ideal guiding the delegates. With a secular faith in the power of dialogue, they were certain that the enormous technological achievements and economic advances of the industrial revolution lay the
groundwork for a new world order. Governments, like persons motivated by self-
interest, would respond to the call for impartial arbitration rather than plunge into
unprofitable violence when disputes arose.

Forty-five new peace organizations formed in the United States alone during
the first decade and a half of the twentieth century. Prominent groups of lawyers,
teachers, and businessmen gave the struggle for peace greater prestige than it had
ever known. In a glorious display of pacifist solidarity, twelve hundred delegates
(including Supreme Court justices, cabinet officers, and other government officials) attended the National Arbitration and Peace Conference, held in New York
in 1907; more than forty thousand persons participated in the Carnegie Hall meetings throughout the event.

Yet behind these vibrant public displays were both ideological and practical
disagreements that would undermine pacifist unity at the dawn of the modern era.
On the ideological level, there was a split between the Socialists and liberals. Although both professed an ideal of internationalism, those on the Left interpreted
war as a necessary entailment of class-riven society where the few profited at the
expense of the many. There could be moments of quiescence during which class
conflict was suppressed, but at heart capitalism remained a system "dripping from
head to toe, from every pore, with blood and dirt."[2] This argument was substanti-
ated by the havoc imperialism wrought in the colonial periphery in the late nineteenth century. Some Socialist pacifists did not altogether eschew violence but,
borrowing from the "just war" tradition, differentiated the inherent and irreme-
dial brutality of capitalism from legitimate, if bloody, revolutionary insurrection.
Proletariat revolution would simultaneously end the class-based violence of human "prehistory" while eliminating the material causes of war between nation-
states.

A second, liberal wing of the pacifist movement was based in the American, British, and French peace societies founded earlier in the nineteenth century. Loosely identified with the emergent bourgeois merchant class living on the seaboard on both sides of the Atlantic and their political representatives, this faction saw in the spread of market capitalism a peaceful future to international relations, one in which "enemies" would be transformed into "competitors."[3] Self-interest unleashed by the market would impel citizens, corporations, and nations to accept mediation, not war, as the means of arbitrating national disputes. Between 1870 and World War I, this movement gained tens of thousands of adherents, many of whom were convinced that the hidden hand within capitalism's global reach had consigned war to the dustbin of history.

This vast bourgeois peace movement was

> a paradoxical political phenomenon, difficult to categorize because the implications of its program were both revolutionary and basically conservative. The changes the pacifists proposed would have revolutionized the system of international politics, substituting law for power as the regulatory principle of international political behavior. Yet pacifists insisted that these changes would not threaten the social or political foundations of existing nation states.[4]

The most significant inconsistency of this liberal wing lay, as with the Socialists, in its uneasy relationship to the "legitimate" use of force. In practical terms, the question of how to enforce decisions made by transnational arbitration remained unsolved. Many pacifists in the early twentieth century held to the need for "civilized" modern nations to retain their arms in order to enforce legally adjudicated decisions in the more "anarchic" regions.

Why, after nearly a century of relative peace and several decades of active promulgation of pacifist ideology, so many millions of young men hastened to the call of arms in 1914 continues to vex students of the period. Perhaps, as the philosopher William James had argued four years earlier, the pacifists of this period misunderstood the nature and the ancient appeal of battle. His article "The Moral Equivalent of War," perhaps the most widely read of any U.S. pacifist tract, challenges one to imagine the conditions for a peaceful society.

As president of the Anti-Imperialist League and fierce polemicist against the U.S. incursion in the Philippines, James had given the matter of warfare a great deal of thought. Arguing that war serves no logical, political, or economic purpose is the error that pacifists of all stripes tend to make. It is a "fact" that conceals a deeper truth. U.S. pacifists in particular, he feared, deluded themselves with their claims that "a world of clerks and teachers, of co-education and zoo-phily, of 'consumers leagues' and 'associated charities,' of industrialism unlimited and feminism unabashed"[5] could replace the appeal and usefulness of battle. They demonstrate a profound underestimation of the attraction of war: "So long as anti-militarists propose no substitute for war's disciplinary function, no moral equivalent of war . . . so long as they fail to realize the *full inwardness of the situation.* And as a rule, they do fail."[6]

Battle is "the supreme theatre of human strenuousness."[7] No other activity affords such a structured opportunity for self-transcendence, or the context (theater) that ensures that acts of valor will be publicly acknowledged. A "moral equivalent of war," for James, must mimic those pristine, unique qualities, virtues that (until now, perhaps) seem only available through combat. Pacifists, with all their calls for arbitration and disarmament, had no categories to measure war's thrilling horror, the élan vital known by soldiers in combat. Until they do, and provide some other, less violent, arena to experience the extremities of loyalty, vigor, and courage found in combat, pacifists will not eliminate battle's compelling grip on our psyche.

James's intuition of a flaccid core at the heart of the vast pacifist movement of his generation offers some partial clues about the patriotic zeal that proved nearly irresistible to one and all with the onset of the Great War. John Keegan, this generation's foremost English military historian, has written:

> By 1914 an entirely unprecedented cultural mood was dom-inating European society, one which accepted the right of the state to demand and the duty of every fit, male individ-ual to render military service, which perceived in the per-formance of military service a necessary training in civic virtue and which rejected the age-old social distinction be-tween the warrior . . . and the rest as an outdated prejudice. . . . Optimism and the moral deprecation of violence could not prevail . . . against the other forces that hurried forward the militarisation of European life.[8]

In 1914 both Socialist and liberal intellectuals and agitators gave up pursuing an equivalence for war and marched into the trenches.

As the war progressed, a chastened, more radical pacifism evolved. In the United States and Europe, feminists, social-gospel clergy, Quakers, and a small group of secular progressive activists bravely resisted the pull toward war. New pacifist organizations such as the Fellowship of Reconciliation (FOR), the Anti-

Enlistment League, the Women's International League for Peace and Freedom (WILPF), and the American Union against Militarism (AUAM) formed during the early years of World War I to struggle against conscription and protest the extraordinary, meaningless carnage of the European "theater."

The United States entered World War I in 1917. An espionage act, passed on 15 June 1917 and amended in May 1918 with even greater restrictions on public speech, proved to be a major obstacle to pacifist organizing and resistance. The act prohibited any form of expression that might "cause or attempt to cause insubordination, disloyalty, mutiny, or refusal of duty in the military or naval forces," or to "willfully obstruct the recruiting or enlistment service." The act hastened the prosecution of almost two thousand persons (not one of them ever convicted as an enemy spy); it "was a singularly successful means of silencing the antiwar opposition."[9]

Subject to new laws that stripped them of their civil liberties, pacifist groups such as the American Union against Militarism found themselves struggling to maintain their right to utter their convictions. This organization, the forerunner of the American Civil Liberties Union (ACLU), reoriented itself toward the preservation of the Bill of Rights in time of war hysteria. The AUAM established a Bureau of Conscientious Objectors, where Lewis Hill would later find employment.[10]

During World War I, four thousand U.S. conscientious objectors (COs) to the draft faced prison terms in harsh and dangerous circumstances, with sentences ranging from twenty to twenty-five years. Portrayed by the government as cowards, the COs were brutally treated by other inmates and often forced to undergo lengthy periods of solitary confinement and other rigors. Roger Baldwin, who helped transform the AUAM into the ACLU, was jailed when he refused to register. In a letter to the draft board, he portrays the ideals of radical pacifism at this time:

> The compelling motive for refusing to comply with the draft is my uncompromising opposition to the principle of conscription of life by the state for any purpose whatever, in time of war or peace. . . . I regard the principle of conscription as a flat contradiction of our cherished ideals of individual freedom, democratic liberty, and Christian teaching.[11]

Baldwin and the ACLU served as an important model for Hill and Pacifica. Baldwin, at times almost single-handedly, blazed the trail that Pacifica, among many, would ultimately follow. The trajectory of the AUAM into the ACLU highlights a core meaning of the First Amendment, which unites expressive and spiritual questing in its language. (That the First Amendment has over the course of the twentieth century come to the fore in nonviolent activism raises an important question about the necessity of liberalism. To resist war in the name of individual conscience, and to promote the pacifist cause under the right of free expression and dissent, presumes a liberal political culture where the claims of conscience are recognized, nonconformity tolerated, if not celebrated, and peaceful arbitration of disputes the norm. The tremendous examples of nonviolent struggle in the twentieth century — until the collapse of Stalinism — from conscientious objection to conscription in England and America and Gandhi's movement in India to the American civil rights, feminist, and antinuclear movements all occurred in nations having a tradition of dissent.)

Those pacifist groups that survived the Great War emerged with great vigor in the twenties. The Women's Peace Society, the FOR, and the WILFP were joined in 1923 by the War Resisters International (WRI). Established to promote the ideals of political pacifism and secular opposition to war, the WRI, and its U.S. affiliate, the War Resisters League (WRL), was, like the ACLU, a vital precursor to Pacifica. In its synthesis of the Socialist critique of capitalism with a Gandhian commitment to nonviolent civil disobedience, the WRI was arguably the first "New Left" organization.[12]

All the postwar social movements in the West were galvanized by the ongoing events in India. Gandhi's accomplishments in putting the mighty British army on the defensive with the Quit India campaign catalyzed a new understanding for pacifists. Gandhi insisted that the "truth force" (or satyagraha) that bonds humans to each other, and to all creation as well, is an impersonal, universal vibration, called ahimsa, or "nonviolence." The greater the intuition of this "force," the more likely it will guide one's action in all areas of existence. (Only contrasted to Christian pacifism — so human-God oriented — does Gandhi's alternative paradigm take on its full, almost awesome, resonance.) A man with millions of "disciples," Gandhi fully understood that unjust authority was never sustained by force alone; it depended on an acquiescent populace lacking a vision of any alternative. With a critique of hegemony as radical as Gramsci's, Gandhi's pacifist movement, massively and against all odds, found a means of generating the qualities of theater, solidarity, and symbolic self-transcendence that James had invoked decades earlier. Gandhi's genius transformed pacifism into a vital, collective "world historical" drama. Pacifist ideology was no longer framed by acquiescent moralizing and individual witness — its basic mode for two thousand years from the Gospels to Tolstoy. Unlike the European and American "pacifists" of the turn of the century, Gandhi linked nonviolence to active resistance to the state, and to Western imperialism more generally. The massive anti-imperialist actions of the Quit India movement stood as proof to pacifists internationally that commitment to nonviolence entailed neither political quietism, nor reverence for the "free" market, nor monotheistic religious trappings. Satyagraha did not lobby government or advocate a change of law; neither did it advocate communist revolution. Rather, the Quit India movement appeared to have discovered a third path, one that tapped into deep traditional sources within the Indian subcontinent while simultaneously transforming some of the basic ideological elements of the caste system. Gandhi recognized "in a very organic way that you had to tackle everything at once, politically, psychologically, economically, socially, religiously. All these things had to be done together."[13] (Although Gandhi was the most influential force shaping twentieth-century pacifism, and global grassroots politics, there is little evidence that persons involved with the origins of Pacifica were ever strict apostles of Gandhi's creed. Nonetheless, they, as millions of others, surely absorbed and acted on the lessons of Gandhi's more assertive approach.)

During the economic turbulence of the 1930s, pacifism was once again an active presence throughout the West. Peace caravans, international institutes, and the Oxford pledge against war all excited the imagination of the public, most especially the youth. At the height of the Great Depression, a spate of books and articles detailed the scale of arms profiteering during World War I. Senate investigations in 1933 revealed that "merchants of death" had lobbied insistently for a "cause" that had injured or killed tens of thousands of American youths and millions worldwide. (It is reported that Woodrow Wilson lamented his actions as commander in chief, stating before he died: "Is there not a man, woman, or child in America — *let me repeat is there any child* — who does not know [World War I] was an industrial and commercial war?")[14] Pulpits throughout the land echoed pacifist sentiments; by 1937, "when Americans were asked 'if another war like the World War develops in Europe, should America take part again?', 95 percent answered 'no.' "[15] The WRL joined the organized Left in linking the increasing turmoil in Europe with the Great Depression; both arose from absence of economic justice within capitalist society.

A number of related international events transformed this bloc of antiwar sentiment. The first was the Spanish civil war. Norman Thomas, a leading pacifist in the FOR, joined the Socialist Party and spoke in favor of armed support of the Spanish republicans: "The Socialist Party position is that it will use the uttermost non-violent methods consistent with true democracy. But . . . it will not yield to fascism anywhere without a struggle and...*nonviolence is not its first and last commandment.*"[16]

The debate over providing material support for the Spanish partisans was but a prelude to a far more intense struggle over the questions of neutrality in the face of Nazi aggression. With no evidence of nonviolent resistance to fascism akin to the Quit India movement, American pacifists found themselves forced into an increasingly difficult corner. On the one hand, armed intervention was repugnant; on the other, fascism must be resisted by any means. The influential minister Reinhold Niebuhr repudiated his previous commitment to pacifism in 1937: "Modern Christian and secular perfectionism, which places a premium upon non-participation in conflict, is a very sentimentalized version of the Christian faith."[17] Revolted by both German and Japanese aggression, and made ever more uncomfortable by their alliances with nativist pro-Hitler groups, the vast majority of pacifists in the United States reluctantly accepted the war effort by the time the Japanese bombed Pearl Harbor.

REVOLUTIONARY NONVIOLENCE IN THE ATOMIC AGE

Nonetheless, more than forty thousand persons eligible for the draft used the conscientious objector provision written in the Selective Service Act of 1940 to refrain from combat. Members of the historic peace churches — Mennonites, Anabaptists, Quakers — were the vast majority of the CO population. Over 90 percent claimed religious sentiment as their motivation.[18]

The government drafted Lewis Hill, a young man from a wealthy Oklahoma oil family, in 1941. Hill, an English major, had recently dropped out of Stanford in his senior year. Deeply held political and philosophical beliefs were an acceptable plea before the draft boards for secular pacifists; Hill was one of the minority who chose a nonreligious creed to justify becoming a conscientious objector. Section 5g of the Selective Service Act of 1941 promised work of "national importance" for those opting out of combat. One hundred fifty-one Civilian Public Service (CPS) camps were established with the goal of providing the country with infrastructure improvements along the lines of the New Deal's Works Progress Administration. Twelve thousand COs opted for the CPS camps rather than serve in noncombatant roles within the armed services. Hill spent some time in a camp in Coleville, California, working on a reclamation project before he was discharged for health reasons. (He was plagued by crippling arthritis for his entire adulthood.) Roy Kepler, a leader of the postwar pacifist movement in the

War Resisters League and a member of Pacifica's first board of directors, claimed:

> My own experience was one of entering the camp in a state of euphoria, having it fortified by the others I met, all these interesting, congenial people that more than made up for the inconveniences. The biggest single mistake the government made was introducing us to each other. . . . They helped build the pacifist network.[19]

In a pioneering work, historian Jim Tracy has recently investigated the transformation of the pacifists during and after their CPS experience, documenting the unique synergy that arose from the pooling of thousands of dissidents in remote and isolated places.[20] During their stay in the camps, the COs not only developed the groundwork for Pacifica but also cultivated the theory and strategy for nonviolent direct action in the United States that would erupt in the civil rights, student, antiwar, ecological, and feminist movements in the coming decades. Tracy concludes his analysis with an important observation: the pacifist vision forged through the camp experience "was a thoroughly American radicalism, for it owed a great deal to the American liberal tradition and the individualist mythology which infused American culture."[21]

This mythology of heroic personal autonomy merged with the practice of participatory democracy in daily camp decisions and led to a greatly heightened fear of native fascism. Hill and other future Pacifica participants easily imagined the result of World War II would be new, excruciating forms of repression, a fear based in part on the massive, vicious Red Scare that followed World War I. Articles such as "America Becoming a Vast Prison Camp" circulated in the CPS newspapers in 1945.

Their response to this projected threat was an enhanced belief in acts of personal protest. Hill would write that the "libertarian revolution requires a widespread urge to refuse authority and to act on a set of principles which might be described as individualist [to ensure] the restitution of the individual in libertarian socialist society."[22] In their camptime isolation, with so many comrades and hours of time, radical pacifists exaggerated the latent totalitarian elements in America. This specter bred an equally fervent, if misplaced, belief that the general public would follow their nonviolent example in laying their bodies on the line against the militarism of the U.S. government.

Thus, as World War II wound to a close, the CO population wavered between the fear that authoritarianism would emerge and the hope that radical social transformation would intervene. Guided by the example of Gandhi and A. J. Muste's calls for "holy disobedience," COs in the camps convinced themselves that their moral witnessing could "raise consciousness in the larger society, leading ulti- mately to a revolution...[beginning with] a spontaneous chain reaction among militant pacifists in the prisons and camps."[23]

Hill, after his medical discharge from camp, emerged as one of the spokesper- sons for this radical tendency when he launched the "Japanese Project" in 1945. Claiming that "only the daring and the fantastic can measure up to the crisis," Hill and a colleague made plans to sail to Japan in the midst of the fighting to sue for peace. Hill explained in a letter to Kepler:

> There were actually some of the most prominent religious pacifists in the country committed (wholeheartedly) to do one of two things: sign a proclamation of peace with the Japanese people along with a public demand on the govern- ment for facilities to communicate it; and/or attempt a sea voyage to Japan (carrying the message) after the govern- ment's refusal.... The plans, very extensive and detailed, included bombarding major cities with airplane leaflets.[24]

While demonstrating the depth of their moral commitment, this plan also serves as an indication of how far they were from accurately gauging the general mood of the public as the war came to a close. A poll in September 1945 indicated that a quarter of the respondents would have "wiped out all Japanese cities at once with atomic bombs"[25] had they been in Truman's place.

With the United States blessed with fantastic geopolitical advantages, most Americans in the aftermath of the war were only too willing to support their na- tion's newfound economic and military might. Guided by zealots such as Dean Acheson and Paul Nitze, Truman led a willing public into the Cold War against the Soviet threat. Only a vast military buildup, based on a nuclear arsenal, could stymie the expansionist Russians. The practical logic for the creation of a nuclear arsenal was cast in Keynesian terms; large government outlays on defense spend- ing, it was claimed,

> would permit, and might be aided by, a build-up of the eco- nomic and military strength of the free world; furthermore, if a dynamic expansion of the economy were achieved, the necessary build-up could be accomplished without a decrease in the national standard of living because the required re- sources could be obtained by siphoning off a part of the an- nual increment in the gross national product.[26]

The synthesis of New Deal pump priming and appeal to national security forged a unique coalition of politicians from across the ideological spectrum who saw in Pax Americana the economic and military security of the free world.

At the end of 1945, a cluster of secular pacifists called a conference to promote "nonviolent revolutionary socialism." These radicals understood before almost anyone else the extent to which the "defensive" posture invoked by the as yet unnamed doctrine of "containment" was the beginning of a new war. At the moment of birth of the national security state, shrouded in atomic secrecy, these hardened veterans of the camps prophesied the permeation of the military into all aspects of American life, from the economy to the family (to say nothing of the aggressive armed intervention in countries around the world). In February 1946, more than one hundred former COs met in Chicago, founding the Committee for Non-Violent Revolution (CNVR). Although Hill had already moved to the Bay Area to promote his radio project and did not attend, he corresponded with the organizers. One letter elaborated his visions for pacifism in distinctly Thoreauvian language, highlighting, once again, individual agency:

> It is one of the great lessons of our time that freedom depends on the responsible disobedience of the individual. When we speak of control of government, we are speaking of the amount of freedom *from* oppressive control which the individual can retain.[27]

The movement for revolutionary nonviolence marshaled personal resources to confront and resist the massive propaganda of the nation-state. However, they lacked, for the most part, the means for publicizing their message and expanding their base. They called for punctual, isolated acts of outrage. Hill, for all his individualist leanings, was suspicious of this overreliance on spontaneity and individual witness. In a letter to Kepler in 1948, about one year before KPFA's first broadcast, he wrote: "The call to action in the absence of a definite idea is really a call for the individual ... to express his anxiety (give it release) instead of compressing it further in an idea."[28]

To remedy this elementary flaw demanded a new approach to both activism and education (and the commercial media), one that Hill had outlined in the initial prospectus for Pacifica:

> A characteristic criticism of [pacifists] rests on their intellectual isolation, their "ivory towerism." There is little doubt that war cannot be prevented primarily through intellectual appeals.... The major job for those determined to see a pa-

cific world in our time is to enter the region close to home, to speak through the newspapers on the street and over the radio stations — in short to identify principles of world understanding where they have direct import in familiar situations.[29]

Certain that citizens went to war in all countries because they were duped by government misinformation, Hill nonetheless held a catholic view about pacifism as a political program. Although he identified with the more radical wing, he was quite conscious of the varied opinions among his colleagues about how best to promote the cause of peace. In this context, he hoped to use radio to test the variety of pacifist orientations in the marketplace of ideas. Radio, in Hill's vision, retained its original potential to transform every living room into a genuine public forum and cultural Mecca, a place where, in the wake of Hiroshima and Auschwitz, people would hearken more seriously to a discussion of peace. This was an astounding idea, given that over the previous two decades corporate broadcasters had consistently eliminated debate of public issues from the airwaves while using culture "to sell toothpaste and soap to the great unwashed."

A man enamored of Kierkegaard and the Western philosophical tradition, Hill's critique of "ivory towerism" was in many ways a self-criticism. In a telling anecdote from his autobiography, the poet Kenneth Rexroth writes about

> the night that Lew Hill . . . showed up unannounced at a large meeting. Lew himself was astonishing . . . a tall thin man with a long, dead white face and a soft, propulsive manner of speaking. . . . He presented what was really a very simple thesis. There had been a great structural change in society, and the days of street meetings and little pamphlets were over. New, far more effective means of communication were available. It was comparatively easy and inexpensive to set up a listener supported FM radio station whose signal would cover at least the entire Bay Area, and which could be supported by the subscriptions without any commercials.
>
> Unfortunately, this was presented not simply, but in the mystifying argot of a professor. It was difficult enough for the younger, college educated people to follow. For the elderly Jews, Italians and Spanish — who after all had been reading revolutionary theory all their lives — it was totally incomprehensible. . . . Perhaps it was rude of me, but I had to act as his interpreter.[30]

> From a philosophical viewpoint, we feel that the health of the
> society may be improved, and cannot possibly be harmed, by
> the existence of at least one radio station that acts, to the best
> of its ability, on significance rather than convenience, and
> actually seeks to serve, rather than exploit.
>
> —Lewis Hill, 1952

Pacifica was not the first manifestly political radio venture. In the mid-1920s, efforts to establish a labor-based station in Chicago bore fruit for a short while. Edward Nockels, secretary of the Chicago Federation of Labor, used organized labor's contacts in Washington, D.C., to receive a license for station WCFL, which he imagined would be the flagship of a "listener-supported, labor broadcasting network [which would] provide a 'working class perspective' on public affairs." In the late 1920s, Nockels consistently argued before Congress and the Federal Radio Commission against the hegemony of corporate media:

> Is it in the public interest, convenience, and necessity that
> all of the ninety channels for radio broadcasting be given to
> capital and its friends and not even one channel to the mil-
> lions that toil? . . . Never in our history has there been such a
> brazen attempt to seize control of the means of communica-
> tion and to dominate public opinion as is now going on in
> the field of radio broadcasting.[1]

Sponsored by voluntary donations from union dues, WCFL was opposed in principle to advertising. With a large listenership in Chicago's working classes, increased by the station's invention of an inexpensive, easily available receiver, WCFL maintained a varied, popular schedule. Contemporary and classical music, major-league baseball, and vaudeville shared airtime with coverage of the strikes and public affairs commentary from a labor perspective. However, the national American Federation of Labor showed little support and refused to use its influence in Washington to guard the WCFL from consistent attacks by the Federal Radio Commission, which changed its frequency, reduced the power of its transmitter, and limited its broadcast time period to daylight hours when its primary audience of workers would be least likely to have the opportunity to listen. Increased operating costs were not met by voluntary subscriptions, leading inexorably to the sale of advertising and the transformation of the station's programming. By the mid-thirties, WCFL's programming was indistinguishable from any commercial station.

There had also been previous attempts to invite the audience to directly subsidize the costs of programming. Perhaps the earliest was Telefon Hirmondo, founded in Hungary in the 1890s. For more than twenty years, this hybrid phenomenon used the telephone lines in Budapest to distribute "a full daily schedule of political, economic, and sporting news, lectures, plays, concerts, and recitations."[2] An estimated six thousand Magyar elite paid a penny a day for the use of this cable service. By 1900 the venture employed over 150 people with a news staff of twelve reporters. Eliminated during World War I, Telefon Hirmondo remains a fascinating precursor of both Pacifica and the contemporary merging of telephone and broadcast industries.

In 1924 a group of Wall Street financiers established the Radio Music Fund Committee to solicit donations from the public to be paid to "artists of the highest calibre" for radio broadcasts at different stations in Manhattan. In Kansas City, WHB, a station owned by Sweeny Auto School, also requested donations from its audience to enhance its cultural offerings, receiving more than $3,000 in response to a written appeal in 1926.

These early attempts at subscriptions never took hold, however, and faded from the public eye. Throughout the late 1920s, broadcasters in this country came to recognize that the most expedient (and profitable) manner of subsidizing programming was through commercialization of the airwaves: the selling of audiences to advertisers. Had Pacifica, through reviving listener sponsorship, merely wrested some of the ether from commercial exploitation, its accomplishment would be worthy of note. However, the foundation incorporated in 1946 had more grandiose plans: promoting the cause of peace and international understanding via the airwaves.

POSTWAR ORIGINS

After leaving the CPS camp at Coleville, California, in 1943, Hill worked with the American Civil Liberties Union in Washington, D.C., counseling draft resisters as head of the National Committee of Conscientious Objectors (NCCO). In 1944 he took a second job working part-time as a news director for WINX, an AM radio station owned by the *Washington Post.* During this period, Hill married Joy Cole, a graduate of the University of Syracuse School of Journalism and Communication and a Washington staff correspondent for *The Nonviolent Activist,* the national pacifist newspaper.

While working at WINX, Hill experienced firsthand the commercial mass media's immense power to frame political and social issues. As Erik Barnouw has detailed, a range of overt and covert "suggestions" flowed from the Office of Censorship in Washington to the broadcast and entertainment industries outlining how the media might assist in the war effort. Unlike World War I, there would be no official government sanctions on the press. Rather, the military and political leaders relied on the voluntary assistance of network officials in producing programs to forge a wartime consensus. This gentleman's agreement paid off handsomely for both sides: the broadcasters received highly favorable tax legislation enabling them to prosper from their record-breaking ad sales; the government in turn received years of consistent, at times overly jingoistic, support for the war effort from the media, a consistent barrage of patriotism that was instrumental in countering the pacifism and isolationism of the late thirties.[3]

It was in this situation that the earliest plans for Pacifica emerged. Hill would later write that radio could be used for the "benevolent propaganda of preventing further warfare through the free and uncensored interchange of the ideas of all civilizations in politics, philosophy, and the arts."[4] Pacifica Foundation was incorporated in California in 1946 to produce radio that would

> engage in any activity that shall contribute to a lasting understanding between nations and between the individuals of all nations, races, creeds, and colors; to gather and disseminate information on the causes of conflict between any and all such groups.[5]

So unique was the project to launch a politically motivated, listener-sponsored radio station in the face of the Cold War and the genesis of contemporary consumer society that it might seem miraculous this effort would still be bearing fruit five decades later. There are at least three interwoven explanations for Pacifica's survival. The first is the visionary leadership of Hill and the tireless devotion he inspired from his early staff, most of whom donated years of their lives for the project. Only twenty-six years old in 1946 when he and his colleagues established the Pacifica Foundation, Hill struck many as a man wise beyond his years — a true renaissance man. As well versed in literature and international politics as with arcane technical matters of broadcast engineering and communication law, Hill was filled with a limitless amount of ambition and vision. Over the years, all these qualities would be necessary to keep the project from crumbling. ("And Lew was the only one who knew how to fix the mimeograph machine when it broke.")[6]

A second explanation for Pacifica's "success" may lie precisely in its humble origins. For almost a decade, KPFA was hardly known outside the Bay Area. Fewer than 100,000 people owned FM receivers in the listening area when the first programs were aired in 1949.[7] KPFA's limited range, and its distance from the centers of power — especially the Federal Communications Commission — meant that the station was sheltered for the first decade from government attacks; these would become ceaseless after the network expanded across the country. Finally, KPFA, and Pacifica more generally, succeeded because Hill's core theory was correct: there were enough people who "recognized their stake" in maintaining a commercial-free media outlet and were willing to sponsor it with annual subscriptions and volunteer energy.

The Bay Area was intentionally chosen as a site for the "brash experiment."[8] With its Mediterranean climate and its cosmopolitan reputation, the region was an appealing place for many "non-conformists."[9] After the war and throughout the following decades, a distinctive political culture prospered in the region. It was the "only place in the country where the New Deal Alliance had not been smashed by red baiting... a special place where people felt free during the McCarthy period."[10] Berkeley English professor Thom Parkinson presents in broad outline the ambience of the Bay Area in the late forties, a milieu that undoubtedly impacted the choice of location for launching KPFA:

> CO's from the war seemed to congregate in San Francisco.
> There they met Italian anarchists of California origin, and
> others with an intellectual interest in anarcho-pacifism....
> With the founders of Pacifica it was their anarcho-pacifist
> belief which motivated them to start the station, with listener-
> sponsorship at the base. It led to the kinds of programs that
> were done, outside of the commercial, outside of the official,
> with a full sense of the play of human freedom. The found-
> ing of KPFA was paralleled by, and in some cases affected
> by, the existence of anarchist discussion groups and com-
> munities throughout the area. Thousands of people would
> attend annual anarchist balls where there was a variety of

> wonderful wine.... McKinney, Hill, Moore, and Triest [the
> original staff and board members of KPFA] attended many
> of these events, often all together.[11]

As the last chapter outlined, the pacifist network emerging from World War II encountered a national and geopolitical situation substantially different from the one that had existed six years earlier. The huge popularity of the war had rendered moot many of the tireless struggles of the thirties, although revelations of the horrors of the concentration camps and nuclear holocaust brought an unprecedented urgency to their cause. The founding of the United Nations and the renewed calls for world federalism provided liberal internationalists with a base for promoting their global vision. Simultaneously the first strains of the Cold War were heard more and more clearly as Truman and the Democrats plied increasingly hawkish rhetoric and policy to shore up the tattered New Deal coalition with the creation of a national security state needing atomic "defense."

Nuclear weapons controlled by the commander in chief fundamentally shifted the constitutional balance of powers that originally granted to Congress alone the authority to declare war. This tectonic slippage in the overall structure of national military policy, and in constitutional government, toward presidential autonomy, was first recognized and opposed by the atomic scientists in their dealings with the military and Presidents Roosevelt and Truman. Their work on the Manhattan Project had spawned a generation of radical scientists who remained throughout the next fifty years among the strongest opponents of the nuclear state. The men and women whom Hill inspired to join him in the Pacifica Foundation shared an important affinity with these dissident scientists. Each operated with the model of communication enunciated by theorists of the time: through changing the opinion of the leaders who set the agenda, they hoped to change the sentiment of the public at large.

Men such as Einstein and Bohr could voice their critique directly to heads of state and high government officials. Their opposition to nuclear proliferation could not stem the tide, however, in no small measure owing to the absence of a popular movement agitating for similar goals. It was toward this end that Hill turned to radio. Much like those scientists who opposed the proliferation of atomic weapons, Hill and his colleagues also aimed their attention at transforming an influential elite. The founders sought "discriminating listeners, addressed as people of intelligence."[12] According to the original plans, Pacifica would appeal to "well defined minorities in the audience of the San Francisco Bay area"[13] by providing "the most thoughtful possible exploration into all issues affecting the individual in each community...in an atmosphere of informality, candor, and freedom."[14] There was no quest for high ratings, but an abiding concern with "thoughtful... exploration" attractive to an influential sector.

When answering the question "What Is an Audience?" Hill stated that "it is not an aggregate of any description, but a single individual. [KPFA] rests completely on this premise and its implications as to the proper genesis of its programs."[15] In another context, Hill explained that "the audience was believed to consist of an individual, whose intention was to listen. The listening individual was assumed to have an alertness, an interest, and an attention-span commensurate with those persons preparing and airing the program."[16]

Was Pacifica elitist; did its initial focus on an educated minority contradict its democratic ethos? This question elicits a variety of answers, both negative and positive, and the next several chapters will return to this theme in various contexts. It is worth recalling that in 1948, "popular" programs in the media were synonymous with the banal and repetitive quality of commercial broadcasts. The very erudition of Pacifica's programming was an implicit, if not overt, comment on the failure of advertisers to support complex programming for those who might enjoy and benefit from more challenging fare.[17] John Downing, refuting the charge that Pacifica's origins demonstrated an implicit elitism, has argued, "The context is everything. In 1949 that context was the continued dominance of rampant commercialism throughout the U.S. airwaves.... It should be realized too that KPFA's internal wage structure was highly radical: everyone was paid the same wage, and decisions were made collectively on all major matters."[18]

Relative to prior efforts within the national pacifist movement, Pacifica was far from elitist. Indeed, it was radio's ability to reach out to a far larger cross section of the population than pamphlets and small newspapers that attracted the initial generation of programmers to the medium in the first place. For the "men and women dedicated to world understanding" who initially worked for Pacifica, radio could extend their influence far more widely than the small newsletters "designed in the main to serve an inner circle of specialized memberships."[19] The original prospectus shows an awareness of the limits of the revolutionary nonviolent movement and the necessity of expanding the base of support for pacifist agitation. For such a project to succeed, there must first be a massive educational campaign, one that should avail itself of the media to expand the pacifist message.

> If prevention of war depends in part upon an overwhelming
> public sentiment against it, the groups most devoted to war

prevention are still without means of widespread communication.... A radio station is unique ... in requiring relatively little initial capital, and returning large earnings in a short period if intelligently operated.[20]

In a twenty-fifth-year retrospective of Pacifica's history, broadcast in 1974, Joy Hill illuminates the political context of the peace movement of the late forties:

> Lew felt the basic commitment which we all needed was to strengthen this country. But we couldn't be strong if we couldn't listen.... This focus on action alone, it seems to me, led to incessant fragmentation. We saw this with the various pacifist groups we worked with. Each one had its own jealously guarded mailing list and essentially they were talking to themselves, continually. We weren't trying to *change* things. We were trying to say, "Hey, man, this is the way it is. Listen to it. If it needs changing, change it, but first you have to know about it."[21]

This confidence in its announcers' sagacity — *"Hey, man, this is the way it is. Listen to it"* — highlights an essential ambiguity within Pacifica's history. On the one hand, preaching to an audience that at times numbered in the hundreds of thousands inflated egos that had little need of enhancement, occasioning bitter internecine struggles. Lew Hill, Elsa Knight Thompson, Ruth Hirschman, Bob Fass — all among the most creative and intelligent radio artisans in the past fifty years — were also forced at various moments to relinquish their positions at Pacifica (or did so "voluntarily" under duress) when their colleagues could no longer tolerate their haughty demeanor. On the other hand, the quest to engage listeners' intellect and passions — to broadcast programs that were occasions for active learning and delight — has encouraged Pacifica's greatest achievements. Before the experiment could succeed, however, a novel form of subsidy was needed.

INVENTING LISTENER SPONSORSHIP

In 1947, the FCC, citing technical problems, refused to grant the foundation its initial license request for an AM station in Richmond, California. This led to a consideration of the relatively unexploited FM band. At that time, very few radio receivers contained FM tuners, and those that did had engineering flaws, making reception tenuous. Nonetheless, it was a fairly straightforward procedure to modify the original license application to request a frequency on the experimental FM band in Berkeley. A letter to that effect was approved by the FCC in March 1948.

A prospectus in mid-1948 soliciting donations for the nonprofit corporation explained that its primary function would be operating a radio station, named KPFA, that would offer news and public affairs, drama, children's programs, and music. The prospectus stressed the opportunity listeners would have to hear live music, given FM's superior fidelity and the station's proximity to university musicians. The first year's budget was targeted at $31,000. Even after the switch to FM, the foundation still believed that "KPFA will be operated commercially and will support itself from commercial revenue after the initial period of stabilization."[22] However, a footnote from this same prospectus indicated that there was

> a plan to eliminate commercial advertising revenue altogether and base support of the station on small annual subscriptions from listeners, who would also receive a monthly publication related to the station's programming. Such a plan, if proven feasible after further study, could be considered only after KPFA has established its audience.[23]

Of all Pacifica's accomplishments, its invention of listener sponsorship remains the linchpin. Listener sponsorship loosed Pacifica from the tether of either corporate or state control. It enabled programmers to experiment with an unparalleled creative freedom of artistic expression while allowing for the ongoing promotion of political alternatives.

A transcript of a program broadcast in 1951, "The Theory of Listener-Sponsored Radio,"[24] contains the closest thing to a manifesto for the Pacifica experiment. As such, it bears some careful consideration. Listener sponsorship is based on two assumptions: "That radio can and should be used for significant communication . . . and that it ought to be subject to the same aesthetic and ethical principles as we apply to any communicative act, including the most personal."[25] Hill argued that the pressure of commercial sponsorship, which impelled producers to attract as wide an audience as possible, accounted "for the mediocrity and exploitation which on the whole signify radio in the United States."[26] According to Hill, a radio announcer on a commercial station succeeds precisely to the extent that he or she is capable of

> simulating emotions, intentions, and beliefs which he does not possess. . . . [There must be] nothing in the announcer's mind except the sound of his voice — no comprehension, no value, no choice, and above all no sense of responsibility. . . . This is the criteria of his job. By suppressing the individual,

the unique, the industry reduces the risk of failure and assures itself a standard product for mass consumption.[27]

In a recent book, James Baughman, describing radio in the late forties, reiterates this point:

> The networks in some ways served as mere brokers between affiliates and advertisers.... The advertiser was sovereign; before airtime, writers of radio comedies made in Hollywood had to teletype their guest star list, their story suggestions, and individual scripts to New York agencies for approval.... Most insisted on the largest possible audience.... With their products so closely associated with individual programs or commentators, advertisers abhorred artistic or political controversy.[28]

Under these circumstances, many prominent intellectuals and artists refused to participate in broadcasting, recognizing that their creative efforts would be manipulated to fit the commercial requirements of the medium.

Hill's insight contains two reciprocal components. The first is that replacing the instrumentality of the "American System" of broadcasting could occur only by transforming the opportunities and motives of the person who

> actually opens his mouth or plays his fiddle.... Even if someone else has decided there should be a broadcast and what should be in it, these are the people that actually make the broadcast.... They constitute most of the radio industry, but are perhaps *the last people we would think of in trying to place the fundamental responsibility for what radio does.*[29]

The second element is the faith that the audience, addressed as intelligent individuals, not as a mass of consumers, would subsidize programming that did not pander to them, even when they could receive the broadcast freely.

Listener sponsorship freed the shackles of the producer-announcer-programmer. On Pacifica,

> people who actually do the broadcasting should be responsible for what and why they broadcast.... The problem [is] not whether you as a listener should choose [only] what you like or agree with—but how to get some genuinely significant choices before you. Radio which aims to do that *must*

express what its practitioners believe to be real, good, beau-
tiful, and so forth, and what they believe is truly at stake in
the assertion of such values.[30]

Inviting the listener to directly subsidize the programming ensures that the announcer (or broadcaster, programmer, producer, all used interchangeably for the
staff who assemble the talent and material that go out on the airwaves) remains
true to the higher calling that Pacifica invokes. However, what if liberating the
staff leads to such idiosyncratic broadcasts that not even "a minority of a minority
gave a hang for their product[?] . . . But it is the reverse possibility that explains
what is most important about listener-sponsorship."[31] Listener sponsorship pro-
duces a "creative tension between broadcaster and audience that constantly reaf-
firms their mutual relevance. Listener-sponsorship . . . inspires a constant exchange
between the staff and the audience, making possible this extremely productive
balance of interest and initiatives."[32]

In this way, the dialogic ideal reenters broadcasting. As a poet and philosopher, Hill never doubted that this ethical reciprocity between staff and audience
was attainable, if only imperfectly. Although measured concretely in ongoing
financial contributions, this relationship between audience-patron and producer
would ultimately be grounded in the ineffable experience of "creative tension"
leading to "mutual relevance."

Hill believed that KPFA would be solvent if 2 percent of its potential audience responded to the challenge of paying ten dollars a year to support the station. After all, he reasoned, a certain number of his listeners would realize that
they are "the only ones who have a real stake in the outcome. . . . Anyone can understand the rationale of listener-sponsorship—unless the station is supported by
those who value it, no one can listen to it, including those who value it." Sub-
scribing to KPFA "implies a kind of 'cultural engagement' some French philoso-
phers call it."[33] (Unbeknownst to Hill in 1951, over time his listeners' *engagement*
would lead them to do more than simply mail annual checks or volunteer to answer phones; they would come to feel they had a mandate to participate in all aspects of station activity, from broadcasting to management policy.)

"THIS IS KPFA"

At the start of 1949, the Pacifica Foundation, consisting of thirteen members from
a variety of pacifist, educational, and broadcasting backgrounds, had raised about
$15,000, primarily from "wealthy Quakers" in the Bay Area. ("None came from

the rich Communist fellow-travelers of which there were dozens.")[34] This was about half of the initial estimate for a bare-bones radio operation. In January, the foundation decided to take the risk and proceed, influenced in no small measure by the Kierkegaardian vision. "Lewis Hill reminded us, 'In crisis — grow. That's the only creative possibility — take a risk and expand.' "[35]

With surplus parts and donated equipment, the staff and volunteers laboriously constructed a station in an office building in downtown Berkeley. The transmitter and most other equipment had been previously used. Large parts of the station's speech-input and audio circuits were custom built. On staff were Lewis Hill, station manager and chairman of the Pacifica Foundation; former NBC employee Eleanor McKinney (who had first encountered Hill while covering the opening of the UN in San Francisco), program director; America Chiarito, music director; and Ed Meece, chief engineer.

In a 1966 broadcast, McKinney recalls the hammers still banging from the laying of the carpets just minutes before the initial broadcast at 3:00 P.M., 15 April 1949.

> Everyone went rushing to their desk and tried to appear to be unconcerned. Lew went on the air saying "This is KPFA. This is our first day on the air." He went on describing something about it.... And of course, the emotional ones among us were full of tears. It was a turning point after three years of working for it. It was quite a day....I don't remember how many hours we were on the air before something went wrong with the transmitter. We had to go off the air while it was being fixed.[36]

The signal, strongest in the Berkeley hills, also reached parts of San Francisco and Oakland and carried almost twenty miles south on the San Francisco peninsula.[37]

KPFA initially broadcast for six hours a day, combining in its schedule public affairs, music, drama, and children's programs. Demonstrating their anarchist proclivities, the staff removed the clocks from the broadcast booths, and "blocks" of program time were eliminated to enable shows, most especially the live musical performances, to flow to their "natural" conclusion. Listeners from this period recall this as a most astounding innovation, so accustomed were they to tightly choreographed network scheduling. Public affairs were intended "to explore minority points of view which seldom receive a hearing on radio and ... emphasi[ze] the basic ethical realities in human relations which underlie all public problems of peace and freedom."[38] Many of the music, drama, and children's programs featured live performances using the talent from the Berkeley area.

A report issued after five months provides much detail about the initial period of operations. The demeanor of the announcers, speaking in their "natural voice," the varied length of programs not truncated to fit arbitrary time slots, and the absence of theme music and other "baseless formalities" all combined to produce

> a radical change in the atmosphere of broadcasting, a renewal of its address to the individual. . . . The Foundation has always contended that far from lacking realism, broadcasting on these principles would answer a great, neglected demand in large areas of the radio audience. What happened during KPFA's first five months on the air proved this contention and the eminent practicality of the station's idealism.[39]

At the heart of the schedule were live broadcasts, both cultural and political. In five months, more than six hundred people found their way to KPFA's microphone. Nearly daily, in-studio musical performances gave "KPFA complete uniqueness in this field, and made it the center of a continuous, area wide music festival."[40]

The station also had

> a large and intensely interested audience for the public affairs broadcasts on controversial subjects — war and peace, race relations, economic democracy — which are the central expressions of Pacifica Foundation's concerns. . . . [T]he subject matter of such broadcasts on KPFA dealt persistently with vital issues of peace and freedom never raised or explored by conventional radio outlets.[41]

Participants on these early shows "would sit for hours afterwards. It was quite exciting to watch people change their point of views during these [post-broadcast] discussions. There was such a great deal of engagement."[42]

The themes of the hour-long public affairs programs ranged from "Conscientious Objectors and the Law," and "Should Labor Form a Third Party?" to the burning question "What Should Be Done about Oleo [margarine]?" An hour devoted each week to "questions of popular culture which led to consideration of fundamental social, economic and philosophical problems" found Seymour Lipset lecturing about "Al Capp and the Shmoo," and roundtables on "The World of the Salesman," "Censorship," and "*Kiss Me Kate* and Shakespeare" among its topics. (Lipset, a commentator until the early sixties, found himself often at odds with the increasingly leftward drift of the station.) A revolving list of individuals, chosen for their ability to present "a radical interpretation of public events, regardless of their particular political or economic doctrine," were allowed fifteen minutes a day to provide a "general interpretation of specific events." The most difficult public affairs show to organize was "The Challenge Table," in which proponents of opposing views were invited to debate each other.

The five-month summary concluded with a succinct overview of Pacifica's founding practices, addressing the question of populism and elitism:

> Any fundamental program of education for peace must commit itself to a continuous traffic in "unpopular" questions. Insights which expose the realities of human relationships in any public situation are always, in their time, "radical." The Foundation's theory has been that the radio medium could be used successfully to educate for peace provided a sufficiently high quality of radio entertainment were [*sic*] provided as the vehicle for educational materials. In short, the "radical" statement of human and societal relations can be effective on radio if placed in a context of general programming which is otherwise highly valuable to listeners.[43]

THE COLD WAR AND DISSENT

There are no tapes in the Pacifica archives that document the public affairs shows from this founding moment. Chronic budget deficits led to the reuse of tapes once they were deemed no longer timely, eliminating the traces of these earliest shows. One of the few surviving programs from slightly later, a roundtable entitled "Is Free Speech Still Free?"[44] dated variously as 1951 or 1952, contains many elements that one imagines characterized the first shows. It is a relaxed conversation between Hill and three panelists: future foundation president Harold Winkler and Lawrence Sears (both educators), and Ray Cope, a minister. Although not an argument for pacifism directly, this program sheds light on the political climate of the time and the difficulty that pacifists and other dissenters faced in publicizing their position.

With the onset of the Cold War, Hill contends, "the ability to engage the right of free speech without fear has been distinctly limited." The public is "not accustomed" to hearing differing sides of controversial issues; hence people are unaware when their freedom of expression is abridged. The issue of free speech is initially framed in terms derived from Alexander Meiklejohn, an adviser to Hill during the station's incorporation. (See chapter 5.) The "privilege of hearing is a precedent condition to making up one's mind wisely in a democracy. It is indispensable for a citizen to act," Hill explains.

Using the contemporary debate over "Who 'Lost' China?" as an example, the panelists concur that there has been an absence of any coherent public discussion on this or other vital issues of foreign policy. Lawrence Sears, president of Mills College, explains, "Let anyone criticize Chiang Kai Chek, and he will find himself in great difficulty. He doesn't have to be a Communist. In so many areas like this people have no chance to hear all sides."

Hill then asks a question that occupies the discussion for the next half hour: "Are we dealing with a problem which is generally the result of international tensions today, or are we talking about a tendency with the collectivization of thought and the reducing of the American heritage?" Harold Winkler, who had recently been fired from his position at Berkeley after refusing to sign a loyalty oath, argues, "With foreign policy carried out on such a grand scale . . . the higher-ups believe there is *the necessity to discipline the American public.* Can a democracy conduct foreign policy . . . with the rapidity of moves that are necessary today? It seems like an insoluble dilemma" (emphasis added).

Although the heavy hand of the state looms over almost all dissident expression, Hill avers that this neither explains nor excuses the pervasive self-censorship on the part of the media and the public at large, an unthinking, generalized acquiescence in the face of the expansive national security state. "Are we Americans concerned with these important issues upon which free speech is being suppressed?" asks Hill. "No, not really," a panelist replies. "The average person is so indoctrinated today from one direction that he doesn't feel the lack of free speech. He doesn't feel curtailed because nothing he thinks is challenging any institution even if he had the courage to speak."

Underlying this apathy and submission to indoctrination is a pervasive "national anxiety." What lies behind this dysfunctional social psychology in which the First Amendment could so easily be abridged? It is "an informal thing," not blatant demagoguery or coercion, but "the fear of the ordinary citizen which lets them just shrug their shoulders and think, 'We'd better get those men [alleged Communists],' something which never would have happened a few decades ago." "Perhaps people do see that this [McCarthyism] is something scurrilous, but they don't move because they don't know what it is that they want to defend. Isn't it a question of not knowing just what values that people absolutely must defend?" Hill asks.

Hill concludes the program by trying to synthesize these various perspectives. The discussion has moved far from its original theme of clear and immediate threats to freedom of speech. Hill is not simply worried that people lack "belief" per se, but — philosopher that he is — he wonders whether, as a nation, citizens can connect their disparate individual sensibilities into any concerted, meaningful action. Repeating the theme of existential alienation, Hill laments that each person has become "an isolate unit in the island society of our time. Can there be any result from our action? How does action have consequence?"

This discussion typifies early Pacifica public affairs programs in several respects. The combination of informality and didacticism was skillfully interwoven; an erudite and often complicated train of ideas was presented casually, with Hill guiding the discourse through many topics. The overlapping themes of dissent, alienation, collective action, and the constraints imposed by the national security state on democratic participation have remained staple concerns of the network until the present.

"Is Free Speech Still Free?" gives an important clue to the station's original ideological sway. Grounded firmly in a libertarian tradition, there is little indication that Hill or any of the original staff had sympathy with the doctrinaire Left. This, of course, could not prevent others from Red-baiting the station. From the start,

> KPFA had been accused of every bias known to political theory, from right wing Republican to Communist. When someone in the first meeting [of listeners in 1950, discussed hereafter] . . . challenged the station with being "extreme left," his remark was greeted with a shout of uproarious laughter by the entire audience.[45]

The founders shared "no exact political or economic credo. But they did share a vital concern with the individual creative way of life in a peaceful society."[46]

At the core of their common hope was using radio to promote the claims of personal conscience. "The conscientious person has an awareness of obligation to himself, to others, and to society. . . . There is a respect which we owe to a human being who is searching for autonomous action, who seeks to build his life on principle."[47] This humanism, deriving from the Quakers and subsequent radical pacifist politics, resonates with the belief that the individual has latent yearnings and capacities only waiting for release. Although sharing much with Deweyan liberalism, Whitman, and certain variants of existentialism, the focus on expressive individualism has but a precarious affinity with the Left, and almost none with the Stalinist inflection of postwar U.S. Communism.

An article celebrating Pacifica's twenty-fifth anniversary in the *Bay Guardian* in 1974 suggested that "the vocal minorities using the station [in the beginning] were generally Communists."[48] Vera Hopkins, the network's archivist, called on Bill Triest, one of Pacifica's original incorporators, to respond to this claim. Triest, an ally of Hill and staff member from 1949 until 1955, wrote Hopkins that

> the station during his years with it was definitely not com-
> munist—*never, never, never.*... all the principles were in
> fact anti-Communist in the sense of being anti-authoritar-
> ian.... They were peace-minded, some were anarchist. Many,
> but not all, were CO's. Their backgrounds were in the AFSC
> (American Friends Service Committee), and War Resisters
> League, not the Communist Party.[49]

To confirm his description of the ideological background of the founders, Triest cites the reading material that was used to fill airtime between shows: Thoreau and articles from *Manas,* a "humanist, pacifist" newsletter from Los Angeles founded and edited by Henry Geiger.

This reference to *Manas* provides a significant clue for specifying the particu-lar democratic radicalism of Pacifica. Founded almost exactly the same moment as KPFA, *Manas*'s first issue in January 1948 championed the importance of Thomas Paine for contemporary America. Paine "gave the love of freedom ... ef-fective direction."[50] He defined liberty in terms the colonists could easily grasp, stirring their passions and providing guidance for the collective action of the Revolution. But what of the present moment? Geiger writes in words that could as easily have been found at the close of "Is Free Speech Still Free?":

> It is as though there were an unspoken cry, lodged in the
> throat of millions: "What shall we believe in? What can we
> work for that will mean something and will last?" ... We have
> to come to grips with the moral realities of our lives, in
> order to have ears to hear what the Paines born to this gen-
> eration may say.[51]

Both Hill and Geiger were pitched toward the same project: coming "to grips with the *moral realities,*" searching for those rabble-rousers able to stir a quies-cent public to action. This ethical idealism unites *Manas* and Pacifica and places both institutions, with Paine, in the tradition of the moral intuitionism originating with the Quakers. Both *Manas* and Pacifica believed that active, engaged dissent unified personal belief and meaningful political action. It was this reading of the

Western ethical tradition (resonating with the Holmesian interpretation of the Constitution as a model of experimentation) that provided the inspiration for these two fledgling projects at the onset of the Cold War.

Given the consistency with which Pacifica attacked the policies of the national security state, one need hardly wonder at the consistent Red-baiting that fills the network's history. If Pacifica actively resisted being pegged as a Marxist or leftist project, it was not, to be sure, out of any sympathy with the bourgeoisie and the regime of postwar American capitalism, but because of the suspicion that the Left, especially the Stalinist variant, had limited, if any, appreciation for the libertarian implications of American democracy.

THE STRUGGLE FOR SURVIVAL

Returning to the organizational dynamics of Pacifica's initial years, one finds that the generous Quakers and other original financial contributors were not enough to sustain the young station. With their ambitious schedule in full operation, Hill, McKinney, and the others were confronted with the overwhelming task of raising the station's weekly budget. The sale of FM radios was falling as the public did not immediately respond to the enhanced fidelity the new technology offered. Without prestigious names to entice potential funders, and with an experiment so unique that it required patient, detailed explanation of its goals and practices, fund-raising proved quite difficult. After fifteen months, having broadcast programs featuring more than two thousand in-studio participants, KPFA had received only 270 subscriptions, most paying between five and ten dollars.

Some other grants and donations supplemented this, but the staff was essentially working voluntarily. At this point, the foundation decided that it was necessary to suspend operations in the summer of 1950 to evaluate the project and determine where to find additional funds. (In discussions of Pacifica's earliest days, the first radio station is sometimes called "KPFA-interim" to highlight the fact of suspended operations.)

At the instigation of perturbed listeners, a community meeting was scheduled in July for the staff and the public to brainstorm ways of keeping the station afloat. The response was overwhelming. Twenty-three hundred dollars were pledged on the spot, and committees of volunteers set up fund-raising and membership drives to hasten the station back to the airwaves. For nine months, KPFA was silent as intense activity continued, with volunteers putting in the equivalent of full-time jobs for the station, sending out fund-raising letters, upgrading the studios, and

helping with general maintenance. In 1951 Hill wrote a letter to his father comment-
ing on the loyalty of the station's supporters:

> The reasons the project is so important to so many people
> are naturally varied. Probably the salient thing is the station's
> function in public affairs programming. In the admissibility
> of heresies and the absence of philosophical censorship there
> is nothing like it in American radio, and we have found that
> the public hunger for a frank speaking out, a radical ethical
> confrontation of major issues, greatly exceeds the public dys-
> pepsia. A more general fact about the project, its complete
> independence, shores up the confidence and respect this kind
> of activity requires.[52]

These first public meetings in the summer of 1950 were momentous occasions, rallying the dispirited staff and proving that the station had indeed accomplished some of its original goals. During the nine months the station was silent,
one thousand listeners subscribed to the station, allowing KPFA to return to the
air in the spring of 1951 with a new, three-kilowatt transmitter and renewed commitment to continue its pioneering venture.

In spite of this support, nagging financial problems persisted. Then, in November 1951, the Fund for Adult Education (FAE) of the Ford Foundation gave
Pacifica a huge grant of $150,000, to be distributed over a three-year period. (A
major donor for a range of cultural and educational activities, the Ford Foundation would continue to play a role in noncommercial media, sponsoring a variety
of experiments in "educational" television in the sixties before the organization
of the Public Broadcasting Service.)[53] This donation was intended to stabilize
station operations and provide more consistent technical resources. Although the
grant enabled a larger fifty-four kilowatt transmitter to be purchased, the station
faced the ironic situation that fewer and fewer FM tuners were being manufactured.
According to McKinney, during the three years of the grant, the national production of FM radios plummeted from 1,400,000 units to 131,000 annually.[54] Had it
not been for the extra subsidy of the grant, it is unlikely that the station would
have survived. It would not be until 1955 that the decline in FM set ownership
halted.

When KPFA returned to the air, the programming was much the same as before. With an expanded three-hour morning slot for children's programming during weekdays, the station broadcast for fifty-seven hours per week. At a time when
more and more artists were being blacklisted, KPFA made it a point to keep its
microphone open to all comers. One particularly poignant moment occurred when

Pete Seeger sang on a birthday tribute for the blacklisted Paul Robeson in the early fifties.[55] During this period, however, problems greater than financing the station arose.

The Pacifica Foundation and KPFA were founded by anarcho-pacifists, bound together by a common worldview and acknowledging Lew Hill as their leader. The license for the station was held by the executive membership of the foundation, numbering around a dozen persons. A committee of directors, a subgroup of six chosen from the executives, elected the foundation officers. Often members of the staff, these officers were also responsible for managing daily operations. In essence, during the first two and a half years of the station's existence, KPFA's staff was the majority of the committee of directors and hence guided both foundation and station policy, following in the main Hill's direction. Hill was both chairman of the foundation and executive director of KPFA. There was also a twenty-five-member advisory council with no decision-making power, composed of prominent members of the community sympathetic with the station's ideals.

This organizational structure began to unravel in late 1951 as members of the original executive membership and station staff moved on to other careers, leaving most power in Hill and McKinney's hands. New members for the foundation were recruited. Although they were scrutinized by the existing board, "they did not fit easily into the near unanimity which had often prevailed in the organization's earlier period."[56]

Some have claimed that the newer members were more generally of the Socialist Left than the original group; another interpretation, offered by Richard Moore, an early staff member and the first officially designated station manager, cites the new members' greater concern with enhancing KPFA's "respectability" in light of the new operating budget.[57] Whatever the underlying causes, the Ford grant and the ensuing new energy notably increased tensions at the station. Ultimately, the problems came to center on Hill's imperious manner. Station personnel almost doubled in 1952; however, as Al Partridge, the program director, later wrote, "people weren't willing to have all the decisions made by a strong man at the top."[58]

Hill's health had been deteriorating, making him less a presence during foundation meetings in the fall of 1952. In September, with Hill on part-time leave, a new position of "station manager" was approved by the executive committee and filled by Richard Moore, a strong Hill ally. Although the committee of directors still held titular control according to the bylaws, many staff and board members believed that the few remaining original staff, often in positions of authority, looked to Hill, Moore, and McKinney to make the key decisions. An observer for the Ford Foundation, called upon to evaluate the status of the grant after the first year, found that chaos reigned. Insulated from impartial and objective oversight, Hill and his "cronies" were not open to working with the new personnel. When the Ford observer suggested staff cuts, the difficulties came out into the open.

The debate to determine whether to cut several part-time staff or one full-time news announcer almost immediately moved to a consideration of Hill's continued position as executive director of the foundation. Should he and a small coterie have de facto control over all station policy, or should the expanded KPFA community of staff and foundation members have ultimate, arguably more democratic, authority? A key question was: "How can one person be both a staff member and a voting member of the policy Board?"[59] A majority of the board argued that the executive committee obtain prior confirmation of the committee of directors for all personnel decisions, a vote signaling "no confidence" in Hill and his partisans who still dominated the smaller executive committee.

As one board member put it, "The important thing is that [Hill] is free to make any investigations, recommendations . . . both at his initiative and by request. But he does not give the orders."[60] For nine months, the station continued to operate while compromise was sought, but none was forthcoming. Finally, in June 1953, Hill submitted his resignation, and much to his surprise, it was accepted. He wrote at that point that he felt himself to be under "more or less continual attack by little minorities, while the majority has never developed any strong belief in . . . my own probity."[61] Complex negotiations once again ensued, which led Hill to feel that Pacifica's democratic ideal, by which all staff were empowered to participate in key decisions, had failed precisely because the majority had refused to trust him: "The rejection and choice of leadership by part [i.e., majority] of the group had, in my own mind, demonstrated that the organizational theory of Pacifica Foundation is unsound."[62] With Hill voted out, several of his closest colleagues also resigned, including McKinney, leaving an entirely new generation at the helm in the fall of 1953, led by Wallace Hamilton.

This new group was able to make good on one of its key promises: increasing subscriptions. From October 1953 until May 1954, more than 600 new listeners joined, bringing the total up to 3,400.[63] However, this coincided with the first full-blown programming crisis. On 22 April 1954, KPFA broadcast a program about the use of marijuana, with four people reputedly smoking the plant in the station studios while arguing for its decriminalization. "Since the summer I turned seventeen, I've become convinced that the only problem the American people have with marijuana (coughing) is that they have a good thing going here and they made it illegal."[64] The uproar in the press was immense; the State Bureau of Narcotics seized the tapes and began an inconclusive investigation.

The pandemonium split the executive committee. Some agreed with Hamilton, who had replaced Hill as executive director, defending the program for its presentation of a controversial viewpoint. Others felt that the lack of "context" for the show, and its blatant sensationalism, caused it to stray far from Pacifica's original, more dignified, vision. Almost the entire advisory board publicly resigned, leading to a dramatic falloff in subscriptions. The Ford Foundation decided to hold up delivery of the final portion of its grant until stability returned.

A meeting of the executive board in August 1954 paved the way for Hill's return by establishing a new position: "President of Pacifica Foundation and Radio Station KPFA." New bylaws established limits on how much input the staff would have in determining foundation matters. The president would have great autonomy over a range of personnel and policy issues. Hill accepted the offer of this new position, leading to the resignation of seven board members.

The letter of resignation of one of these members, a Hamilton supporter, vividly depicts how deep the passions ran:

> It is hard to convey the pleasure I experience at finally leaving Pacifica. In all my life I have never seen so many fanatic disciples and emotional women gathered in one organization. And what is most amazing is the way they pretend to the highest ethical principles while engaging in the most vicious Machiavellian activities. Black is white, (if Lew Hill says so), and words take on new (and opposite) meanings. Reorganization means kicking out your enemies.... Healing process means getting Lewis Hill back.... For the past year we have had control of Pacifica. We began to make decisions on the basis of objectivity as each of us understood it instead of on political or ideological grounds. But we had to spend much of our time fighting off those who were determined to bring Lew Hill back by whatever means.[65]

Immediately upon Hill's return to control in the fall of 1954, the last part of the Ford grant was released. In early 1955, a milestone was reached: the number of subscribers surpassed 2 percent of the FM receivers in the Bay Area, the benchmark of Hill's theory of listener sponsorship. Yet even with this success, the station was not self-supporting. A fund-raising promotion by Hill broadcast at this time speaks of the station's urgent need to receive additional donations:

> The station is obliged to operate with about one-third of its cost not covered by subscriptions.... This deficit must be made up. 6,500 dollars in accrued obligations hangs over

the station in a most threatening manner. . . . Do not imagine that this problem is merely a theatrical affair going in a Chicago studio of a network where all problems are solved by dramaturgy. If KPFA is part of an important community in which you feel yourself a participant, the problem is as real and immediate to your interests as, shall I say, the monthly bills you pay, for services you could not otherwise receive.[66]

These chronic financial problems led to internecine squabbles throughout 1956 and 1957. Although the subscription levels continued to climb, reaching almost five thousand by the spring of 1957, operating costs grew as well. An article from this period in the station's monthly newsletter, the *Folio,* explains that

personnel is the chief expense on KPFA's budget. . . . Entering employment at KPFA involves neither a vow of poverty, nor a vow of celibacy, and since growing families are dependent on the station's payroll it has been necessary to institute a living wage.[67]

New plans in 1957 called for wealthier listeners to pledge donations above the base subscription rate of ten dollars a year, but these funds were increasingly difficult to solicit. By July, the debt had reached almost eighteen thousand dollars.[68]

During the spring, several members of the board's executive committee once again attempted to oust Hill, and in response, Hill fired the executive director. More significantly, in June 1957, Hill dismissed two staff members, both of whom also belonged to the California Federation of Teachers. Their response was to file a grievance with the station, and also to protest their firing to the union. The union in turn announced that it considered calling a strike at the station, although just what jurisdiction it actually had to do so remained unclear. Hill insisted that as president of the foundation, he retained the control of all personnel matters and was opposed to the grievance committee's recommendation that severance pay be extended to diffuse the threat of a strike. Nonetheless, the committee of directors overrode Hill's wishes and accepted the grievance committee's report.

The next day, Hill drove his 1953 Dodge to a nearby hill and committed suicide using carbon monoxide poisoning, leaving only a brief note:

Not for anger or despair
but for peace and a kind of home.

He was thirty-eight years old.

Robert Schutz, the foundation's executive director who had been fired in April, speculated a decade later that no one of us was big enough to show Hill that we cared for him — no one in that station — no one in his life really could convince him that he was cared for. He was an utterly lonely man. And he isolated himself by his sharp tongue, his wit, and his intelligence.[69]

During the last few years of Hill's life, his arthritis caused him increasing pain, preventing him at times from rising from a chair without assistance. To alleviate the pain, his dose of cortisone was gradually increased, likely exacerbating his moodiness. Strains in his marriage during the last year of his life had also led to a brief separation from his wife. But of all the psychological pressures, the most weighty was his shouldering the huge burden of keeping his radio station alive.

In October, Harold Winkler succeeded Hill as president of Pacifica and was appointed manager of KPFA. Between the time of Hill's suicide and Winkler's election, thirty thousand dollars were contributed to the foundation, providing operating funds for the remainder of the year. In the following April, KPFA was selected to receive the George Foster Peabody Award for Public Service, one of broadcasting's highest accolades, for "courageous venture into the lightly trafficked field of thoughtful broadcasting, and for demonstration that mature entertainment plus ideas constitute public service broadcasting at its best."[70]

4. The Development of the Pacifica Network

> Every year of [KPFA's] life there has been internecine warfare
> here. There is a very simple explanation for this. The people
> who work at the station are individuals with strong minds and
> strong points of view. That's why they're here in the first
> place. . . . The station programming is for people with strong
> opinions. It would be a sad day indeed at KPFA if things are
> comfortable. . . . I can foresee no future time when KPFA will
> be smugly sitting here on a nice income with everyone feeling
> satisfied saying, "We've made it, fellas, now we can just coast
> along." . . .
>
> We're still a brawling, vigorous, active, impassioned bunch
> of people working here. And as long as you have this kind of
> people involved at this station, you'll never have a nice,
> comfortable sitting-back kind of feeling. We attack, and we are
> attacked all the time.
>
> —Al Partridge, KPFA manager, "KPFA's Sixteenth
> Birthday," KPFA, Pacifica Foundation,
> 15 April 1965

Early Ambitions and Expansion

A 1948 prospectus seeking donations for KPFA promised donors that "after an initial period of stabilization," commercial revenue would support the station. More than that, "its income will eventually create a surplus providing for its own expansion or the establishment of other stations."[1] Although these business plans did not come to pass precisely in the manner imagined, airing Pacifica's signals and ideals beyond the Bay Area remained dear to the hearts of many in the foundation.

The earliest goals of Pacifica had anticipated that the foundation's enterprises would include a bookstore, a publishing house, and other media outlets.[2] One project, long under discussion but never to be realized, was the publication of a monthly literary journal modeled on the BBC's *Listener*. Articles, poetry, transcripts of programs, and a monthly broadcast schedule would reach a national audience as well as serve Pacifica's Bay Area subscribers. With advertisements on a quarter of the journal's proposed seventy-two pages, this self-supporting venture would grow to be independent of the radio station.

Hill had also kept close contacts with colleagues in educational television and indubitably maintained visions of expanding into that realm as well. In 1951 he was central in the formation of the Bay Area Educational Television Association, which was responsible for the newly licensed "educational" station KQED. The thought that noncommercial television would follow in the path of educational radio appalled Hill. He wrote a colleague in 1951 that

> the first problem in developing educational TV is to get it completely separated from the history and organization of "educational" broadcasting. We must face the fact that the main use of university radio stations has not been to form a cultural bridge between centers of learning and occupational classes.... There is no evidence that these stations and their organizations (NAEB, etc.) even understand the basic functional obstacles to development of new art forms.... Moreover, the people in charge of educational stations are tied either to state legislatures or to boards of trustees which inevitably represent tendencies close to the commercial and conservative part of the community.... The real poets and musicians are working like hell at every conceivable job, and most of the best political thinkers are not in politics. These people are continually producing the real stuff of our century.[3]

Hill closes this letter with a call for permanent government subsidy for noncommercial television, overseen by an independent committee, with income supplemented by audience subscribers — in other words, a program very similar to the one ultimately recommended by the 1967 Carnegie Commission on public broadcasting that launched the Public Broadcasting Service (PBS).

Hill's preoccupation with KPFA limited the amount of time he could spend developing plans for educational television. Nonetheless, his vision of noncommercial listener-sponsored television has had a lasting legacy. San Francisco's KQED public radio and television outlets hired many of the staff and volunteers from KPFA.[4] Its fund-raising efforts, modeled on Hill's principles of listener sponsorship, would be one of the central paradigms for PBS a decade later. Nonetheless, the expansive system of public broadcasting that emerged in the late 1960s became far more dependent on government and corporate backing than Pacifica, accounting in the main for its more timid approach to programming.

The grant from the Ford Foundation in 1951 led to an aborted effort to establish an AM outlet for KPFA in the Bay Area. Several years later, Pacifica's first successful expansion project got off the ground: a plan to expand listener-sponsored broadcasting to Southern California. A budget and report from early 1957 detailed the prospects for a midstate link capable of relaying the programming from Berkeley to a new Pacifica station in Los Angeles. A small amount of local production in the Los Angeles area would give that station "its own distinct identity within Pacifica's signal system, and it would be enabled to grow as rapidly toward a larger staff and independent scheduling as the development of Southern California subscriptions permitted." Original estimates projected the need for about 2,700 subscribers in Southern California during the first year to make this second outlet viable.[5]

The midstate transmitter was never built, but the hope for a second station in Southern California persisted. An enlarged prospectus was published in July 1957, the time of Hill's suicide. This document made more explicit the proposals of the network's expansion. Instead of producing the bulk of programming in Northern California, the plan now suggested a semiautonomous station in Los Angeles, with increased local production. Reiterating KPFA's ideals of providing "an arena for the calm consideration of ideas, [where] given intelligence, freedom and a wide assortment of ideas to work from, creative people will be stimulated to *think*," the prospectus spelled out the highlights of the Berkeley station's history.[6] Although the overall financial situation in Berkeley at the time was far from opulent, there was a definite optimism to the tone of the report: "The crucial demonstration in KPFA is the fact that this kind of broadcasting, which has never before emerged in America, can now be developed and supported by a method which assures its integrity as well as its life."[7]

After Hill's suicide, foundation vice president William Webb worked in Los Angeles to raise funds and generate community support for the new station. The official construction permit application for a Los Angeles station was sent to the FCC in January 1958, requesting that Pacifica be allowed to broadcast at 90.7 with a 47,800-watt transmitter. Two other colleges had also put in requests for that frequency, reputedly at the behest of commercial enterprises that wanted to keep the band space open until they could use it. Webb successfully negotiated this problem by convincing the other institutions to rescind their requests; the FCC granted Pacifica its second frequency in December 1958.

During this period, organizational concerns arose over the autonomy of the new outlet, christened KPFK. Webb was hoping for KPFK to be fully independent of its sister station; Pacifica attorney Harry Flotkin argued that the licensing arrangement in which the foundation was the licensee for both made this legally impossible. In this dispute, Webb ultimately resigned, but only after he had raised enough money via grants and subscriptions to acquire a transmitter for the station.[8]

Terry Drinkwater took over station management responsibilities after Webb's resignation; KPFK went on the air in July 1959 with predominantly local programming. Daily mailings of tapes between the two California stations established an affiliation between them. With a powerful signal heard from Santa Barbara in the north to the Mexican border in the south, KPFK immediately had a much larger potential audience than its sister station. The bulk of KPFK's initial schedule stressed classical, modern, and folk music, although public affairs, drama and literature, and children's shows were also broadcast. By and large, it would have been quite difficult to distinguish between the overall programming of the two Pacifica stations in the first several years of KPFK's operations.

The addition of a new station necessitated certain organizational changes within the foundation. The committee of directors was enlarged from eleven to fifteen and then to twenty-one members. An executive committee held certain power to oversee problems in the much larger region in which the foundation now operated. A local Los Angeles board, composed of the area's "left-liberal elite," helped govern their own station KPFK's policies. Soon members of this local board were active in foundation politics; one member, Lloyd Smith, was elected chair of the national board of directors in 1964.

"Highbrow's Delight"

In 1959 philanthropist Louis Schweitzer called the main office at KPFA and offered to donate his New York radio station, WBAI, to the Pacifica network. It is rumored that whoever answered the phone simply hung up, believing Schweitzer to be a crank.[9] On the third attempt, the utterly serious if somewhat eccentric New York millionaire was able to speak directly with foundation president Harold Winkler. After a brief conversation, he was able to convince Winkler of his credentials. (Another account offers that Schweitzer was ready to take back the offer, believing that no one at the foundation took him seriously.)[10]

Having purchased the commercial station in the mid-fifties as "a hobby," Schweitzer at the time intended to make WBAI "a kind of Off Broadway radio station."[11] He personally covered the deficits the station ran. Then, during a 1958 newspaper strike, the station found sponsors clamoring for airtime. Said Schweitzer:

> We had more commercials than we could handle. And I listened to the station and thought it was awful. From the commercial point of view, we were being most successful, and I

realized right then that was not what I wanted at all. I saw
that if the station ever succeeded, it would be a failure.[12]

He put the station up for sale but refused all offers because the purchasers were
"not interested in maintaining the type of program [he] had aimed at." At that
point, he concluded

> that if I wanted this station to succeed, I had an easy way
> out. All I had to do was to give it away to Pacifica.... Now
> and then, WBAI had rebroadcast tapes of programs from
> KPFA, and I admired what they were doing out there. They
> were doing exactly what I would have liked to do. So I picked
> up the phone, called Mr. Winkler out in Berkeley, and told
> him that if Pacifica would like to have a station in New York,
> I would be happy to give him one.[13]

WBAI went on the air as a Pacifica affiliate in January 1960. Like KPFK, it adopted
an unquestionably "Pacifica" format (called "Highbrow's Delight" in a headline
from an article on Pacifica in *Time* magazine).[14] Classical music predominated,
combined with extensive public affairs ("The Social Role of the Dentist") and
political commentary. Programs questioning civil defense and lambasting the new
hawkish Democratic administration ("A Radical Shift in Nuclear Policy")[15] were
among the specials of WBAI's first year.

The three stations in the network now had a potential daily audience of almost
twenty million listeners. They

> each were free to work out their own programming in light
> of their own concept of service to their respective communi-
> ties. But they also draw freely on each other's resources,
> forming a network without wires of educational communi-
> cation across the country.[16]

A *New York Times* editorial summed up the sound of the network:

> The Pacifica stations are frankly esoteric, even a little pre-
> cious in their music, outspoken and often controversial in
> their discussion programs. Their standard offering, in con-
> trast to that of the popular music stations which blanket the
> country, is symphonies and symposiums.[17]

In an era when radio formats were becoming increasingly rigid, the eclectic, "serious" contemporary musical performances that filled Pacifica's schedule were

unique and garnered abiding audience loyalty. For dozens of composers, writers, and commentators, Pacifica was the only outlet for their work; and for tens, if not hundreds, of thousands of listeners, the network provided the only alternative to three-minute news summaries, pop music, and old masters' symphonies offered by commercially driven radio.

There is no indication that Hill, or anyone else in the station's early history, worried that the complexity of much of the programming was inherently elite. Hill had written in 1952:

> KPFA does not think of itself as "highbrow," split from the huge, real mass of the population. Rather, the station finds itself in the middle of that war which is splitting public opinion between those who believe that man has important choices to make and those who are cynical or fanatic about human chances. If KPFA is "with" anything, it is clearly on the side of those who believe in the mind and who endeavor to make responsible mentality more socially effective. It has never been felt at the station that the height of the listener's brow is a tenth as important as what goes on underneath it.[18]

Hill and other programmers knew that there were many urbane liberals who, while not willing to commit civil disobedience or profess a revolutionary critique of society, might be influenced "to make responsible mentality more socially effective." Indeed, so fully had he moved beyond his youthful rhetorical flourishes that those who knew Hill only in his work with Pacifica called him "a New Deal liberal" or a "social reformer," hardly the characterization of the young anarcho-pacifist sailing to Japan with visions of a "libertarian socialist society."

The *Times* claim that Pacifica was "frankly esoteric" invites consideration of how the programs themselves reflected on the question of class, culture, and taste. A two-hour roundtable discussion with Dwight MacDonald, Daniel Bell, and Winston White addressed this question of elitism and the meaning of culture, democracy, and American mass society. (Hill, with many of the pacifists of the postwar decade, looked to MacDonald as one of their ideological leaders: "It would be impossible to overestimate the influence of *Politics* [MacDonald's journal] on Lew in those days.")[19]

This freewheeling conversation grappled with two interchangeable issues — whether contemporary popular culture was aesthetically inferior to earlier folk entertainments such as bear-baiting, and whether "elite" culture had a home in modern "mass" society.[20] Winston White, the most ecumenical of the three, suggests that the "crisis" felt by academics and other critics surrounding the "quality" of contemporary culture is overblown. He makes the positively radical suggestion, so resonant of contemporary theory, that soap operas and popular novels enable wide audiences to consider and respond to important ethical and emotional problems in a way not altogether unlike earlier audience reactions to Shakespeare.

This populism infuriates MacDonald, who throughout the discussion adopts the most explicitly mandarin position. He bemoans the fate of "excellence" in the United States, where a "professionalization" and "eclecticism" have infected all elements of high-cultural production. For an example, he cites the American literary magazines — "crude, tasteless, filled with articles written by second-rate college professors." Bell responds by defending the pluralism of the United States without championing "mass culture," which he accuses of being an inchoate combination of too many influences.

Perhaps, the moderator suggests, the harsh assessment of mass culture is an implicit condemnation of the working class. This comment elicits a flurry of comments from MacDonald, who claims that there "really is no working class in the United States, only middle class and people who want to be middle class" to Bell's peroration that there is no condescension whatsoever in pointing out that the bulk of mass culture is "obviously trivial, hideous, dispiriting, lacking in taste." Bell follows this with an important observation: to the extent that intellectuals do lambaste popular culture, in the main they refuse to engage with any of the mass media directly. "Except for a very few outlets like this station, what are the alternatives, what does one do about it? Why not insist that a certain number of [television] channels are set aside purely for serious productions?"

Bell, attacking MacDonald, then presents a concise outline of the epistemic "problem" that mass culture raises for critics of all stripes. Standards of aesthetic quality

> have been replaced by categories. What contemporary sociology has done is to establish categories like "high brow," "middle-brow," "lowbrow," or "mass-cult"/"mid-cult," and then, based on the audience reaction, or on the presumed intention of the work, judges these works on the basis of the categories rather than on explicit literary or aesthetic merit. It seems to me what happens is that popular sociology has replaced the critical function of determining what is good and bad.

The result, according to Bell, is a false empiricism that judges the work by the conditions under which it was produced, or the audience for whom it is intended, but with "no critical evaluation" of the work itself. "This is both bad for sociology as well as for literary and aesthetic standards."

Bell and MacDonald continue to debate, at times heatedly, about the heuristic value of categories such as "mass-cult" and "mid-cult." Bell contends that in modern consumer society, it is impossible to separate the consumption and the production in the realm of culture. "The good" and "the popular" have become increasingly indistinct, leading to the evisceration of disinterested judgments of quality. In spite of moments of honest disputation, there is an underlying affinity among the three commentators: they all agree on the inherent superiority of such "objectively excellent" works from Shakespeare, Proust, Eliot, and Stravinsky.

One fascinating example of how controversy over the politicization of aesthetic categories comes into play in Pacifica's history occurred in KPFK's first years. There was an intense dispute in Los Angeles in the early 1960s over the programming of "genuine" folk music. KPFK began broadcasting at the emergence of the "folk revival." Its eclectic acoustic programs have been among the most long lasting and popular on KPFK's schedule, building an intensely loyal audience during the early sixties and maintaining a significant listenership for decades afterward. But the very popularity of these programs in the early days provided the occasion for questioning to what degree commercial success and gold records were antithetical to real "folkorica."

> For example, the Kingston Trio was absolutely *out.* . . . There was all this pounding on the table and discussion going on and on. . . . KPFK reflected that passionate division between people who wanted genuine folk music, and those who wanted something broader. KPFK was always a purist.[21]

The conundrum that this raises is stark: unadulterated "folk" music can be judged as either an "oppositional" category to commercialized bowdlerizations such as the Kingston Trio or, conversely, background music for liberal aficionados living in Bel Aire. In its purity, this music can represent a means for the people, the "folk," to maintain a form of expression beyond, or beneath, that produced for the market. On the other hand, the peculiar devotion of an urbane, wealthy audience for "authentic folk music" can easily be portrayed as an elitist — "highbrow" — gesture, an attempt to preempt the broadcasting of new, potentially popular performances that do not conform to the established paradigms of what certain programmers and listeners consider pure "folk." (Think, in this regard, of Bob Dylan's first electric performances, where the audience called him "Judas.")

During the early sixties, this debate was generally carried out at a fairly civilized level. While the issues raised were substantive and revealed complex differences in approaches to cultural politics, the actual audience for whom the issues resonated was relatively small; the "folk" themselves, the vast population of sec-ond-generation Midwestern "dust bowl refugees" inhabiting the Los Angeles hinterlands, was not the main target of KPFK's programs. More important, the stations had not yet entered a political epoch in which questions of musical taste would reveal the stands one would take on other, more consequential matters.

INTO THE MAELSTROM

A two-hour documentary produced at WBAI in 1961, "After the Silent Genera-tion," indicates the central political concerns animating the network as the sixties began: three students from the Student Non-Violent Coordinating Committee (SNCC) discuss the sit-ins in South Carolina; a young beatnik describes what it was like moving from the South to the Mecca of Provincetown, Massachusetts; a Dartmouth professor in New Hampshire mobilizes his neighbors against the jail-ing of a clergyman indicted for refusing to answer questions before the state Un-American Affairs committee; California students protest against capital punish-ment; a woman from Brooklyn College rallies the campus to a march against nuclear weapons; high-school students disrupt their graduation to show their con-tempt for a new, strict principal; and representatives from a new organization, "Students for a Democratic Society," discuss their organizing principles. "While the goals which they all express may not be related—indeed most do not—they all do have one thing in common. All these young people are in revolt against some aspect of American society. They are *protesting*."[22] "After the Silent Gener-ation" heralds the

> thawing of McCarthyism on campuses around the country, with youth taking their place at the vanguard of progress. . . . They're out on the picket line, acting in solidarity with the students in the South. Altogether, they are searching, sitting up and taking notice. Youth is questioning the wisdom of its elders.

The opening segment on the civil rights movement is the most complex, mix-ing a dozen interviews with activists and bystanders recorded in many cases on the street. Driven by a "passion for freedom and human dignity for all people, re-gardless," a young woman from SNCC invokes the spiritual and political center of the movement, which lies as much in the nonviolence training as in the actual sit-ins (or "tests") themselves. Building trust during the training is central to the political education. As the students describe in some detail their belief in nonviolence, an older black man chuckles with delight. These young people have broken through the complacency of his contemporaries, who until now had "not believed that any change was possible."

A student then describes in very subdued tones her experience of being arrested in a Tallahassee sit-in. Police were "amazed to be around college-educated Negroes. They were used to drunks." Why, she is asked, do the youth put themselves on the line? Is she concerned by the lack of support from the older generation? She hesitates, then responds softly, perhaps thinking of her own parents, "Everyone wants equality, but the young Negroes are just beginning to live. The older people have adjusted themselves, and we have not."

The students from SNCC are clearly the elder statesmen of the activists in this program. They have by 1961 already been through years of resistance struggle, subject to beatings, jailings, and worse. In 1961 the direct action movement led by students in the civil rights struggle (themselves educated by the radical pacifists of World War II) served as a vibrant indication of the shifting political climate. The civil rights movement has "stimulated young people everywhere. New campus groups and political organizations are springing up everywhere." (Throughout the decade, Pacifica programs on the free speech movement and the antiwar, feminist, and antinuclear movements consistently refer to SNCC as the central model for nonviolent struggle.)

On this program, the rebellious high-school students seem to be out of place; theirs was an incident where some kids let off end-of-year steam at graduation a bit too rambunctiously. The analyses offered in the other segments are more complex. In each case, students or youth more generally are shown to be at the forefront in the protest against a variety of social problems. The most articulate and politically sophisticated are the activists from the Students for a Democratic Society.

Al Haber, Andre Schiffrin, and Carol Wisebrod speak with passion about a number of social problems. For these students, the question of political transformation is complex. They contend that the student movement is radical only "compared to 'the nothing' that has been going on before.... Whether students have come to radically new perspective in their role in society, I doubt." They claim to be unimpressed by the various protest movements against the bomb and against civil defense, neither of which has the stature of the civil rights movement. Echoing Hill's earlier argument about the tactics of the Committee for Non-Violent Revolution in 1946, Haber asserts that many protests are simply "personal responses to social problems," not a coherent, strategic confrontation with the system itself.

Schiffrin expresses in germinal form the understanding that will shortly inform *The Port Huron Statement,* anticipating the leftward movement of students in the coming decade. He insists that the sort of

> Quaker witnessing activities against the nuclear bomb are not political. It's just what happens when students feel that all conventional forms of protest have been shut.... The two-party system waters down demands, for example, for nationalized health care.... Most leading politicians, commentators, and intellectuals have given to the young people the attitude that there are no issues, no problems, everyone's prosperous; they have obscured all the many problems that exist, questions of genuine equality, of the work most people have to do, of overall degeneration of the community and the lack of communal life. The result is that when a college student does discover an issue it comes to him as a complete surprise.... His response is to personalize it and his action is often naive.

This critique of naïveté, remedied by exposing the systemic nature of social ills, is a harbinger of the increasing radicalization of SDS's program.

The program ends with a critical overview by Margaret Mead. Her commentary is fascinating in its ambivalent support of the young people. Although she praises the idealism and activism of the youth, she also argues emphatically, in words that would be echoed by Nixon a decade later, that just because someone "chooses to protest doesn't make that person a patriot.... Students are indulged to a tremendous degree. Never before have students been given such great publicity." In a harsh critique most likely aimed at the high-school graduation protest, she claims that some of the students act "just like hoodlums and think they should be treated like heroes."

Mead contends that the protests against the bomb and against capital punishment are examples of American democracy at work. The very act of dissent, not the substance of the issues involved, proves that

> we're not living under Communism, and have the privilege of expressing ourselves.... What we need to do in a free society, and in a world where there is more communication than has ever existed in the human race, is to have a spectrum. If every student is radical, then there would be this great conformity....

> Very few demonstrations have been blown up to enor-
> mous proportions by the media. This is fine with what hap-
> pened with the students in the South, but it is difficult living
> within this glare of publicity. All the first steps and plan-
> ning are blared out into everyone's living room.

Such publicity cannot ultimately benefit the organizers, Mead contends.

ENTER THE FEDS

As is widely known, no good deed goes unpunished. The early sixties marked the
first serious attempts at government surveillance and outright suppression of Pacifica. The 5 February 1960 issue of *Counterattack*, the weekly newsletter dedicated to providing "facts to combat Communism and those who aid its causes,"
headlined "Radio Station Promotes Communists."[23] This newsletter, founded in
1947 by the American Business Consultants (the group that had put out the infa-
mous *Red Channels* in 1951, launching the Red Scare in Hollywood), cataloged
Pacifica's injudicious decision to program "radicals" such as Norman Cousins,
Carey McWilliams, and Edgar Snow, as well as actual Communists such as W. E. B.
Du Bois and Herbert Aptheker. Most suspicious of all was listener sponsorship, a
"method of operation so unusual as to be revolutionary itself."[24] This lingering
McCarthyism would not normally have deserved much attention had not it been
shortly followed by congressional attacks making very similar charges.

Over the course of its first five decades, Pacifica has been scrutinized by the
Senate Internal Security Subcommittee (SISS), the FCC, and ultimately the Supreme Court. None of these threats have ever borne fruit in outright censorship
or license revocation and have in some cases inspired the defense of the network
by those who had not listened to it or supported it in the past. Nonetheless, the
network's ongoing battles with the government have not been won without cost.
The next chapter will consider more generally the issue of the First Amendment and its centrality for Pacifica's operations. Here the subject is the first sustained effort of government interference in Pacifica's operations, the 1962–1963
Senate Internal Security Subcommittee investigation into Communist infiltration
of Pacifica.

Hill had written in the forties that Communists resembled commercial broadcasters; both were unable to think and speak for themselves, mouthing a script
formed elsewhere.[25] There is no evidence that the principal founders ever considered themselves, or the Pacifica project, aligned with any organized leftist group.
Nonetheless, the libertarian principles that undergird the programming philoso-

phy led to widely diverse commentators, some of them self-identified Communist Party members: Dorothy Healey, longtime California organizer (who for decades hosted shows on various Pacifica stations), Steve Murdock, writer for the party newspaper *People's World,* and the historian Herbert Aptheker, among others. William Mandel, a commentator and Soviet sympathizer (active for more than thirty years at KPFA), was also identified by the Senate investigators as "linked to several Communist Front organizations."[26] Another charge leveled by the Senate concerned the fact that WBAI had broadcast some of the Communist Party's 1959 convention, as well as some taped programs from Radio Moscow.

The key target for investigation was Jerry Shore, a vice president of the foundation, slated to assume the presidency in 1963. Shore's reputed political history, and that of his ex-wife, would become the center of ongoing controversy for both the government and the network. Before joining Pacifica, Shore had worked with John Lewis and the CIO and was known to be an active left labor organizer, a fact that he had never denied. His radical background was well known when he had been nominated for the board by a San Francisco minister who had heard Shore speak at his Yale Divinity School graduation in the late thirties.[27]

The political affiliation of these personalities, one might suggest, was less significant in the speed and avidity with which Connecticut senator Thomas Dodd called the SISS together to scrutinize the network than the content of a certain electrifying broadcast that aired on WBAI on 18 October 1962. Dodd convened the special inquisition into Pacifica's affairs within three months of WBAI's exposé of the FBI's illegal activities by former special agent Jack Levine. "The first sustained attack on the FBI and its director, J. Edgar Hoover, ever presented by American radio or television"[28] aroused intense, immediate government concern. The Justice Department, hearing that WBAI was soon to broadcast the former agent's "confession," suggested to the station through intermediaries that airing the program might not "be in the public's interest."[29] However, since all the allegations were true, and Levine was no longer an employee of the agency, there was no legal way to prevent him from speaking.

Hoover was well aware of Levine's apostasy. As the broadcast date approached, he "placed the station under siege."[30] WBAI offered the FBI time to respond, but they refused. In the days leading up to the broadcast, key station personnel were harassed at home, a wife of a programmer was threatened with arrest, and restaurants where employees ate received bomb threats. An editorial in a New York paper called for the agency to "lay off the station."[31]

In this pathbreaking interview, the former special agent described in detail the manner in which the bureau infiltrated CORE, NAACP, ACLU, and other liberal organizations. In some cells of the Communist Party, Levine claimed that FBI agents actually were the majority of the dues-paying members. Phones were tapped, mail opened, offices and homes bugged. Although all these facts subsequently became widely known, Levine was

> the first defector from the FBI, at a time when no journalist in America would dare criticize Hoover. The FBI was above reproach. . . . Levine was the first person to crack that facade. He had gone to the *New York Times,* the *Washington Post,* all the major networks, the BBC for all I know, and nobody would touch him with a ten-foot pole. . . . Then he was told to come down to Pacifica, "they'll run anything."[32]

To protect itself, the station made numerous tapes of the interview available to news organizations.

Once broadcast, the program was a rousing success. Covered by journalists from around the country, and later the basis for a book by investigative reporter Fred Cook, the revelations brought unprecedented notoriety for Pacifica. The phones at WBAI were tied up for hours after the station went off the air at eleven o'clock; more than one thousand dollars in donations were pledged that evening. (Eight years later, KPFA would broadcast an exposé of the COINTELPRO operations, one of the first public discussions of the continuing abuses of the FBI.)

The repercussions, however, were swift. Although none of the bomb threats, arrests, or other reprisals ever came to pass, the fact that the SISS so eagerly leaped to investigate the network on spurious charges (some up to five years old) on the heels of the broadcast struck participants at the time as far from coincidental. As Robert Brustein wrote in the *New Republic* in the following year, "I am suggesting, in short, the possibility of bureaucratic conspiracy. In the case of Pacifica, this is more than a possibility."[33] In December 1962, subpoenas to appear at a Senate hearing were mailed out to key network personnel, who found out about the summonses from a press leak at the *San Francisco Chronicle* before they arrived in the mail.

Over the course of the two-week hearing in Washington, beginning on 10 January 1963, seven witnesses were called, including former Reed College president and foundation board member Peter Odegard, Acting President Trevor Thomas, Dorothy Healey, and Shore. Requests that the hearing be open to the public and broadcast live were denied. Compounding Pacifica's difficulties during the hearings was the fact that all of its three licenses were either up for renewal or pending authorization (in the case of the transfer of WBAI) by the FCC. There was

debate among the witnesses concerning whether Pacifica should choose to participate or refuse on the basis of First Amendment privileges. As Trevor Thomas
wrote, explaining why he and the others did testify:

> We might question the propriety of this legislative hearing.
> However, we choose not to do so, not because we concede
> the right to compel testimony in these matters, but because
> Pacifica's policies and programs are open to everyone. No
> subpoena is necessary to secure this kind of information.
> Those responsible for making and administering Pacifica's
> policies will discuss these policies and our programs with
> anyone, including the Senators. We also, of course, respect
> the rights of an individual compelled to speak under sub-
> poena to respond to purely personal questions in the light of
> his own conscience and understanding of his constitutional
> rights.[34]

The case against Pacifica made by principal investigator Senator Dodd was
chimerical from the start. After acknowledging that the government has no right
to interfere with the content of the programming itself, Dodd claims that the
hearings are based on the fact that

> our world has become so vast and complex that the average
> person is completely dependent upon mass communication
> media for his knowledge of the outside world. Communist
> control over these media would present the gravest threat to
> our national security. Any substantial Communist infiltra-
> tion of these media, which would give influence to agents
> of a foreign totalitarian power seeking to poison the well-
> springs of public opinion in the United States, would be of
> concern to this Subcommittee.[35]

According to Dodd, the investigation of Pacifica was carried out "for the protection of the freedom of the press." If indeed it were true that Communists had
slyly taken over Pacifica without warning, they would be able to broadcast "un-
true and scurrilous matter, without the consent of those nominally in charge."[36]

Peter Odegard eloquently responded to these charges.

> During fourteen years Pacifica Foundation has, I believe,
> lived up to the highest standards of broadcasting and has
> earned the confidence and respect of tens of thousands of
> listeners.... Pacifica Foundation believes that the American

people are entitled to have not only the best in radio entertainment and education, but also to have access to the full
spectrum of ideas from right to left. Unlike Communists or
Fascists [or, he might have added, the SISS], Pacifica is not
afraid to let the people know, nor afraid to let the people
judge.

To operate stations of this kind is not easy. They are bad-
gered and beset, criticized and denounced by both left and
right. There is scarcely a program with which someone does
not disagree. Nevertheless, I believe that Pacifica Foundation offers a radio program service that is both unique and
unexcelled anywhere in this country or abroad. And, Mr.
Chairman, may I finally say with all deference and respect,
that if your committee by summoning us here under com-
pulsory process impugns the loyalty or casts doubt upon the
dedication of these stations to the highest ideals of this republic, you may destroy us. And if you do, I believe you
will do a disservice to the cause of freedom to which I hope
both you and the Pacifica Foundation are committed.[37]

The government struggled throughout the case to focus on the personalities
involved, trying to argue that individuals in positions of responsibility and high
profile had disguised their Communist affiliations. As the *New York Times* edito-
rialized, under

the committee's bizarre criterion, [it] has the right to inquire
into the background and beliefs of everyone working for an
organ of public opinion.... The inquiry looks like an attempt to make Pacifica conform to a concept of speech that
is pleasing rather than free.[38]

No discussion of Pacifica's programs or overall policy was ever entered into the
record. As the trial continued, it was clear that the government really had no
"case" at all against the network; it was essentially mimicking the allegations, in
some cases almost verbatim, brought two years earlier by *Counterattack.* In the
end, no charges or sanctions were levied.

There were two repercussions of the great publicity surrounding the SISS hearings. On the one hand, dozens of newspapers, from the *New York Times* to the *Los
Angeles Jewish Voice,* wrote editorials supporting Pacifica. S. I. Hayakawa, then
a professor of linguistics in Berkeley, helped found Friends of Free Radio, which
placed full-page ads in major newspapers and encouraged civic leaders around

the country to speak out in support of Pacifica. In January 1963, KPFA alone received more than one thousand new subscriptions and renewals; KPFK and WBAI, more than six hundred each. For the most part, listener loyalty proved to be deep and abiding.

However, between the end of the hearings and the granting of the permanent licenses to WBAI and KPFK, the board was impelled to confront the issue of loyalty oaths. On 7 October 1963, the FCC asked the board, the foundation officers, and the station managers to sign under oath a questionnaire regarding current or past Communist affiliation. The basis for this request is found in Section 308 (b) of the 1934 Communications Act, which stated that

> All applications for station licenses, or modifications or renewals thereof, shall set forth such facts as the Commission by regulation may prescribe as to the citizenship, character, and financial, technical, and other qualifications of the applicant to operate the station ... and other such information as it may require.[39]

There was no certainty that refusal to sign the oath would entail the loss of licenses, but the omens pointed in that direction. Not signing would give the commission the opportunity of sidestepping the issue of loyalty and Communism, basing its refusal of licenses solely on the fact of noncooperation with a sanctioned request.

This FCC inquisition occasioned months of intense, at times vituperative, internal debate. Pacifica's lawyers claimed that outright refusal and potential litigation would cost over one hundred thousand dollars, with only minimal guarantee of success in light of earlier court decisions upholding the right of the FCC to require loyalty oaths.[40] The overarching issue was whether any sort of consensual action was possible among those affected by the directive. The staff of all three stations were united in their adamant desire to refuse any participation. The members of the board of directors were equivocal, with some who had already signed such documents in the past not in a position to make a firm stand on principle, and others claiming it would be better to sacrifice the licenses (and hence the entire Pacifica experiment) if it meant acceding to the FCC's demand. Others felt that the foundation needed new lawyers who would be more willing to take the commission on. The policy that had developed in light of the SISS hearing was to allow each person the freedom to determine individually how to handle such requests. However, the blanket nature of the FCC's letter seemed to undermine this possibility.

By November 1963, the press had been alerted to the controversy and once again lined up solidly behind Pacifica. The *New York Times* wrote another strong editorial praising the network:

> After all the noble statements about wastelands and the need for more education and controversial programs, is the FCC going to nullify its preachments by a witch-hunting approach to stations that offer a place on the airwaves for unpopular views? Here is a splendid chance for the new FCC chairman, E. William Henry, to give some meaning to the usual platitudes about freedom of the air.[41]

The "resolution" to this crisis was a statement drafted and agreed to by the board that upheld the fact that

> the officers, directors, and station managers of Pacifica clearly and unequivocally affirm their support of the Constitution of the United States, both in spirit and in formal terms. We believe that these affirmations, coupled with the record of operation of the Pacifica stations, establish the qualification of the Foundation to operate in the public interest.[42]

During meetings held in mid-November, the commission seemed to look favorably on this compromise, but not without holding open its right to individually question key personnel if necessary. This seemed a veiled threat toward foundation vice president Shore, who had become the KPFK general manager by that time as well. He told the board that he would refuse any cooperation with the FCC and subsequently resigned, citing health reasons, although he readily admits the board forced him out to avoid further conflicts with the FCC.[43]

In January 1964, the FCC granted all the licenses, stating emphatically that the commission favorably viewed the sort of programming Pacifica offered. In a speech several months later before the National Association of Broadcasters, E. William Henry, chair of the FCC, lashed out at the organization for their blasé attitude during the entire episode:

> When a regulatory agency is called upon to deal with allegedly obscene Communists on the airwaves, it has a hot potato on its hands. The Commission cleared Pacifica of all charges leveled against it. . . . Now at any public meeting of broadcasters I have attended, a speaker only needs to make reference to freedom, or heavy handed bureaucrats, and he will receive applause. But oratory is easy and action is more

difficult. Surely in Pacifica's case, if ever there was a time when freedom of broadcasting was at issue, this was it. And I ask you, who took the action? Which association sent delegations to Congress? Which of you wrote me a letter demanding that the Commission dismiss the charges? . . . Where were your libertarian lawyers and their *amicus* briefs, your industry statesmen and their ringing speeches? If broadcasters or their advocates felt involved in this issue, there is no evidence to indicate those views. Not one commercial station felt obliged to make its views known to the Commission. . . . Your contrasting reaction to these two struggles [FCC investigation into overcommercialization and the delay in relicensing Pacifica Foundation], in my judgment casts a disturbing light on the basic motivations of an industry licensed to do business in the public interest. . . .

When you display more interest in your freedom to suffocate the public with commercials than in upholding your freedom to provide provocative variety, when you cry censorship and call for faith in the Founding Fathers' wisdom only to protect your balance sheet, when you remain silent in the face of a threat that could shake the First Amendment proud oak to its very roots, you tarnish the ideals enshrined in the Constitution and invite an attitude of suspicion.[44]

Shore's forced "resignation" infuriated the staff at all three stations, which went silent for a period in January to protest the board's policy. Louis Schweitzer, who because of FCC inaction still retained titular control over WBAI, and who had befriended Shore, made an offer to transfer the station to him directly instead of to Pacifica. Shore refused this generosity.[45]

ELSA KNIGHT THOMPSON

The staff displeasure lingered. Almost all programmers intensely disputed the board's advice to sign the loyalty oaths. Spearheading the charge against the board was KPFA Public Affairs director Elsa Knight Thompson, after Hill the most important and controversial figure in Pacifica's history.

Hired by Hill in 1957 to sell ads for KPFA's program guide, the *Folio,* Thompson soon showed her skills as an incisive interviewer whose tenacity and political acumen would serve as a model for an entire generation of Pacifica reporters. An American citizen married to a British diplomat stationed in Romania in the thirties, Thompson headed the BBC's service of broadcasting to occupied Europe during World War II. She was not a pacifist by inclination, and her politics have been described as "democratic socialist" and "[British] Labour Party" radical.[46] Like Hill, she believed that radio must combine the best in cosmopolitan culture with public affairs and news programming devoted to international understanding. Chris Koch, the reporter who had interviewed former FBI agent Levine at WBAI and was the first American reporter to visit Hanoi during the Vietnam War (and subsequently helped create NPR's *All Things Considered*), was one of Thompson's first protégés. He has called Thompson Hill's "spiritual heir."[47]

As were all main figures in Pacifica's history, Thompson was passionately devoted to free speech, which for her meant a commitment to broadcast only that which a person believed to be true: "You have to have both a consciousness and a conscience about society, or you shouldn't be on the air."[48] She insisted that the recognition and expression of the "truth" was possible only if a person took a particular moral stance. This inevitably entailed disputation, not consensus; as program director, she strove to include as many different perspectives as were available. While it was never easy to include ongoing commentary from conservative perspectives, Thompson made a strong effort, interviewing Nazis and allowing right-wing spokespersons from such organizations as the John Birch Society open access to Pacifica's microphone, leading to what some have called the "freest commentary service ever heard on radio."[49]

In the late fifties, Thompson personally produced programs on the sociologist C. Wright Mills, "homosexuals and society," the end of the Hollywood blacklist, and other controversial topics. A stickler for grammatical and technical excellence, she demanded complete control over all broadcasts in her department, expecting to preview tapes at least a week before their broadcast date. None of her shows from the late fifties, however, was as dramatic as a three-hour special report on the riots that took place during the 1960 House Un-American Affairs Committee (HUAC) meeting in San Francisco. More than two hundred students, joined by longshoremen and others, protested these hearings on 12 to 14 May. Tapes from inside the chamber where the hearings took place, including biting testimony from KPFA commentator William Mandel, were spliced with actuality from the protests themselves, in which police used fire hoses to blast the protesters. The recordings reveal how the ensuing pandemonium overwhelmed the reporters on-site. The HUAC documentary was an "amalgam of different tapes which were left on at this time," spliced together to provide a narrative coherence, a sort of radio vérité.[50]

Like the radio drama of the thirties, this complex aural environment placed the listener directly in the center of the events. This collage style would come to be used on Pacifica more often in the early sixties, not only for news and public affairs but also for experimental documentaries on the "hipster" scene in Venice, California, or nightlife in San Francisco. This experiment in ambient auditory environments attempted to recreate on radio an overlap of sounds that could "resemble a film from Godard."[51]

Three years later, in the wake of the FCC loyalty oath controversy, Thompson led a united front of department heads from all three stations in opposition to the board's equivocating stance. These dissidents insisted that the board refuse any cooperation with the FCC. They were furious that foundation president Jorgensen and board chairman Trevor Thomas had pressured Shore to resign. Thompson told KPFA's promotion director that a "weak and spineless Board had caved in to the FCC," following a pattern begun in 1960 when President Harold Winkler had discontinued Herbert Aptheker's commentary series in face of right-wing complaints.[52] Then Thompson had circulated a letter claiming that Winkler "had seriously undermined [the staff's] faith in his guidance of Pacifica, and the democratic principles along which staff members have always heretofore proceeded."[53] She caucused with staff, volunteer workers, and department heads constantly, leading one member to believe that she "hindered their work by the psychological atmosphere and the long hours of discussion. . . . I can't recall her ever suggesting that we just sit down with the manager and talk over what was bothering us."[54]

After Pacifica's episode with the FCC in early 1964, Thompson coordinated a restructuring proposal for all stations that would give more authority to the department heads, diminishing the overall power of the station managers. In her outline, no independent producer could work outside the departmental structure. Thompson's proposal in some ways simply reflected the smaller role that the station managers had played in the daily affairs of the stations over the past year, under the burden of supplying documents and depositions in the struggles with Congress and the FCC. Although widely supported by most of the staff in all three stations, her proposal soon became a major bone of contention with the managers and the board. Much as in the debate about Hill almost a decade earlier, the arguments about policy were inseparable from issues of personality.

Some members of the board claimed that Thompson clove to a narrowly defined orthodoxy (vaguely leftist, although clearly nonpartisan) while refusing to brook any criticism of her management style or programming. Board president Jorgensen continued this criticism by writing that Thompson had become "an increasingly poor interviewer because she sought to discredit persons she did not agree with."[55] There is little evidence bearing this assessment out. Although Thompson surely had an imperious style, shaping every detail of the public affairs programming to her liking, even those who recall her with some ambivalence maintain she was never less than a consummate journalist. Among Thompson's most demonstrable flaws was what one coworker saw as an inability to work with women, toward whom "she behaved detestably."[56] Others who worked with her believed that her insistence on clearing any public affairs program before it aired created both a formal and a tacit aura of censorship.[57]

Exacerbating these problems was the emerging tension between the paid staff and the volunteers. Although the problem was not caused only by Thompson, her fierce protection of her favorites and cold shoulder to those with whom she did not wish to work epitomized a growing schism among regular staff and part-time, generally unpaid, programmers. In the struggles over this issue, KPFA lost one of its most prominent and well-known volunteers, film critic Pauline Kael, who had been airing her opinions on film and other cultural topics for years. Kael tried to bring the issue of discrimination against volunteers into the open and was in turn vilified by staff and listeners for publicizing internal problems. In one of Kael's final broadcasts, she reveals the depth of this problem:

> Liberals always talk about pressures to conformity as if they were addressing a problem "out there." But there is a KPFA kind of conformity too. If you're ever going to do anything about "out there," you had better start right here. . . . I have no power at the station; I've never been asked for my suggestions; I've never been invited to any staff meetings. Some of us who have been on the air for years have never even met each other. I don't know any more about who sets KPFA policy than I did when I first started working here in 1954. The typical KPFA listener regards criticism of the station the same way that a member of the chamber of commerce regards an attack on the profit system. "Go back where you came from," they tell me.[58]

Kael left Pacifica soon after this 1962 broadcast, but the problems she addressed would linger.

Despite Jorgensen's criticisms, Thompson's interviews with activists and leaders from the civil rights movement, the ACLU, and the peace movement continued. An interview with Student Nonviolent Coordinating Committee leader James Farmer a month after his historical freedom ride through the South shows her ability to inspire her guests to self-reflection and insight. In this program, she leaps directly to issues of strategy: How does nonviolence work to reconstruct society? What are its limitations? Where does Farmer believe the movement to be heading?[59] Thompson's listeners received a frank and comprehensive overview of the civil rights struggle and its directions for the future, not just a recapitulation of the momentary horrors of the bus attacks certain to provoke liberal guilt. In examples such as this, one can easily comprehend how inspirational Thompson was for the younger news staff—the way her style of committed journalism provided a transition into the sixties.

We might recall the opening statements from the program discussed earlier, "After the Silent Generation." There Mark Starr spoke of the ferment of the youth, now once again "occupying itself in the vanguard of progress."[60] It is clear that Thompson wished to position KPFA, and Pacifica more generally, as part of this movement. Board members such as Russell Jorgensen, who had left his position at the American Friends Service Committee to become Pacifica's president, remained true to their understanding of the network's pacifist and liberal humanitarian heritage, but they also believed that a "journalist" must remain a neutral observer to objectively report events; Thompson and the dozens of young male reporters she trained were actively promoting a bolder, more engaged form of radio that participated within and promoted the movements it covered.

Issues came to a head at KPFA in the winter of 1964 when Jorgensen and KPFA station manager Trevor Thomas decided to fire Thompson for disrupting station activities. In March, the board simultaneously sent letters confirming Thompson's dismissal and explaining that the board would not recognize a new staff union until a vote of all employees could be held. This action riled the staff, who were inclined to see Thompson's firing as related to her activities in union organizing, a contention that the board vehemently denied.[61] On 23 March 1964, a strike was called, shutting the station down for three hours until volunteers and nonstriking staff resumed operations. Demanding Thompson's immediate reinstatement, the union organizers finally accepted arbitration and a fifteen-minute daily time slot for them to broadcast their version of events. During this time, Thompson herself was hospitalized and did not participate in the swirl of controversy that surrounded her firing.

In May an agreement with the union was reached, ending the threat of a strike. Thompson's case went before an arbitration committee six months later, and a mixed finding was ultimately released in July 1965. The committee found that Thompson had been fired without just cause and deserved to be reinstated in her job with partial financial recompense; nonetheless, it did not exonerate her altogether, claiming that "Elsa's actions substantially contributed toward her firing and that a resumption of such conduct, if it occurs, will be an appropriate ground for discharge."[62] According to Pacifica archivist Vera Hopkins, the union was able to suppress this portion of the arbitration notice and treated the verdict as a triumph.

Thompson once described her approach to her public affairs programming as the assemblage of as many different commentators as possible in order to provide the audience with a range of perspectives on the complex and evolving nature of social reality. She scathingly contrasted this with the static "liberal" metaphysics of broadcast objectivity in which "if you can somehow find the middle of things, and then just squat there, virtue has been achieved."[63]

The sort of "balancing" that Pacifica did present was consistent in its critique of the status quo. The programs during Thompson's regime and later indeed represented a vast spectrum of opinions, but almost all perspectives accepted the premise that racism, American imperialism, patriarchy, and greed were forces of evil. For example, from 1968 to 1971, there were hundreds of programs in which speakers such as Eldridge Cleaver or Abbie Hoffman forecast the incipient revolution, challenging the listeners to prepare themselves and join, or "at least get out of the way"[64] when the action began. As a counterpoint, Saul Alinsky and Herbert Marcuse and later a range of feminists and other activists from the new social movements — no less harsh in their assessment of social problems, the Vietnam War, and government policies — warned against this infantile and macho posturing. Alinsky in particular broadcast a blistering critique in 1969 against what he denounced as the "complete idiocy"[65] of the Weather Underground and its fetishization of violence.

Thus, as the sixties developed, the stations moved with their staff and audience to the left. Yet unlike many of the more radical movements of the decade, they rarely fell prey to a simplistic or doctrinaire analysis of imperialism or capitalist society. For every exhortation to cleave uncritically to the wisdom of Lenin or Mao that found its way to the microphone, there would be a program or series hosted by Paul Sweezy, Lewis Mumford, or Paul Goodman. These figures, while maintaining the same vehemence of critique of U.S. society, would cast it in a far more sophisticated (dialectical) light.

THE TURBULENT DECADE

Two practices dominated the formal production of the public affairs programming of the sixties. Many shows provided open microphones for important, and often-lesser known, leaders of the student, antiwar, civil rights, Native American, and feminist movements. For instance, Mike Klonsky, past president of the SDS, could use KPFA as a megaphone urging students at all grade levels to leave school in 1969 in preparation for the revolutionary youth movement. In a dialogue from a Barnard forum on homosexuality, an older lesbian expressed the pain she felt whenever her younger compatriots used the term "dyke" with pride — a word she felt resonated with derision. Eldridge Cleaver and Jerry Rubin shared a podium in Berkeley in September 1968, analyzing the upcoming presidential election in which the Yippies were running a pig, Pigasus, for the office of commander in chief.

In these programs, and hundreds like them, volunteers at the different stations were given a cassette recorder or reel-to-reel deck and told to cover a speech, panel, protest, or rally. The close surveillance of Elsa Knight Thompson or Hill over the audio productions gave way to a radio vérité format in which the microphone was left on and the resulting tape played back essentially unedited. Perhaps 15 percent of the archive tapes are this sort of program, representing an immense repository of unexpurgated "actuality" from that period.

However, the influence of Hill and Thompson remained a significant force, as evidenced by the many meticulously produced, complex documentaries from this era. One in particular, a four-part series from 1971 entitled *The Turbulent Decade,* reveals how Pacifica's growing body of archival material could be used by skilled and committed broadcast journalists.

Divided into segments entitled "Militarism and Democracy," "The Civil Rights and Black Power Movements," "The Counter Culture and the Antiwar Movement," and "Violence in the United States," this series sits like a bookend with "After the Silent Generation." In the later series, Dale Minor, Chris Koch, and other producers present a thorough and panoramic overview of the different currents that produced such social turbulence. "Militarism and Democracy," the first and most elaborate of the four broadcasts, demonstrates the incipient fear of militarism voiced in the program "Is Free Speech Still Free?" in the early fifties to have been no idle speculation. Clips from Eisenhower, Kennedy, and Johnson, to Andreas Papandreou, the Greek leader removed by the U.S.-backed junta, all paint a complex picture of the antidemocratic forces impelling U.S. imperialism around the world. Vietnam, while the most important focus, is only the most bloody of many interventions that the United States had sponsored over the past generation — from Iran and Guatemala to Greece — during which time information about arms sales and military assistance was carefully shielded from the American public. Seymour Melman and other academic critics discuss the imbrication of the

U.S. political economy with Pentagon spending, chastising the forces in the government that foist weapons on South American and African countries to prop up the military budget. The segment expertly traces the manner with which the United States' best political and intellectual efforts were channeled into Pentagon activities.[66]

Each of the hour-long programs in this series is dramatically structured and densely packed with interviews, commentary, and actuality, demonstrating a deep commitment to the craft of audio production. More than a synopsis of a decade, the series also documents the tragic denouement of some of the most important movements of the sixties, especially those of the civil rights groups and students. The tapes reveal how these vibrant forces for social transformation became blinded by their own revolutionary rhetoric, with their participants finding themselves increasingly marginalized and on the defensive during the height of Nixon's presidency.

No elder commentator for this series serves the role Margaret Mead played in the earlier tape. However, two years later, a broadcast lecture by famed educator Robert Hutchins serves much the same purpose. In his discussion of the sixties and the youth rebellion, Hutchins proves himself generally more sympathetic than Mead to the causes of the young. He mercilessly attacks the university, the Pentagon, Watergate, and the stagnating economy; students and the underclass are following the only logical course open to them when they rise up against a social system that is crumbling at every point. As Hutchins argues, when "every institution in society was revealed to be failing," the youth have "no option except revolution."[67]

INNOVATIONS

Throughout the sixties, Pacifica stations developed according both to the local conditions of the areas in which they operated and in conjunction with the national and international political ferment. Interestingly, at no station did rock music make extensive inroads; during the height of San Francisco's psychedelic era, KPFA had no more than one or two weekly programs devoted to "popular" music. An apocryphal tale has it that a WBAI announcer was fired for playing occasional rock and roll as filler between programs, against station policy. Nonetheless, if the network as a whole was slow to adopt pop music as part of its schedule, it did innovate in a number of other ways.

KPFA was among the first stations in the Bay Area to broadcast on the FM band. Each station pioneered the extensive use of the technology enabling callers to speak live to a host over the air. At first, in the early 1960s, these call-in shows were slotted in unused, late-evening blocks when the management saw little risk in allowing random volunteers a chance at the microphone. Soon, however, KPFK and WBAI had developed large, devoted audiences for this novel experiment in dialogue. KPFA's use of call-ins remained more "dignified," with William Mandel hosting on-the-air discussions of international affairs and Soviet politics in a scholarly forum. His program seemed far more in line with traditional Pacifica concerns compared with the increasingly anarchic and fluid shows emerging at the sister stations.

A recent anecdotal survey of the emergence of talk radio speaks of this format expressing

> the essence of culture in the most commonplace exchanges.... Talk radio is a potentially democratic medium. [But] the show is not a vehicle of unadulterated pure speech, but of orchestrated pure speech, a fusion, as it were, of provocation, censorship, and self-expression.[68]

At Pacifica, there was far less censorship, as provocation and self-expression of both host and caller dominated. For example, on *Radio Unnameable,* WBAI's midnight-to-whenever open mike hosted by Bob Fass, the discussion could go on for weeks about the reputedly hallucinogenic properties of banana peels, with biochemists and postdoctorate pharmacologists offering extremely detailed scientific information between calls from people who claimed to have, just before phoning, experimented on themselves. Fass also encouraged far more serious discussions centering on police brutality, undercover "narcs" in high schools, and Vietnam.

In Los Angeles, *Radio Free Oz,* first airing in 1966, was a call-in cum "theater of the air" hosted by Paul Robbins and Peter Bergman, who would later form Firesign Theater.

> As a team they spoke to, and on behalf of, a segment of our community that had never been addressed on such a scale before. It is hard to define their audience ... young in spirit and resistant to the society that spawned them.[69]

The first regularly scheduled "youth"-oriented show on KPFK, *Radio Free Oz* aired every night at midnight so as not to offend the station's traditional listener-sponsors. It was more elaborately theatrical and willfully manic than any previous KPFK production, with sound effects and improvisational skits building to pandemonious climaxes, followed immediately by cerebral and quite serious discussions of Zen metaphysics and the power of love.

The contemporary controversial topics—sex, drugs, the war—that filled KPFK's and WBAI's early-morning airwaves were also dealt with on commercial stations' call-in shows during this period. However, on commercial stations, the host and management faced the imperative of maintaining a close watch on the dialogue, using such mechanisms as the tape delay and the "panic button," which were available but rarely used at Pacifica. For commercial stations, call-in radio's spontaneity is

> dangerous. To survive talk radio had somehow to contain rather than eliminate its perilous unpredictability. . . . [In the 1960s] the talk show successfully commodified disruption, turning controversy, politics, and "shocking alternatives" into consumables.[70]

This containment strategy was for the most part foreign to Pacifica. To be sure, there was a certain pandering quality to some of the discussions of sex and drugs, but listeners rarely felt that the conversations were circumscribed to meet commercial imperatives or prurient fears.

There were several important results of these new after-midnight shows. The first was the unique, if ineffable, bond they knit between the stations and the emerging phenomenon of "the Sixties." The daytime schedules at all the stations remained much within the "highbrow" paradigm throughout the decade (with the exception of WBAI, under the extraordinary influence of Fass—see chapter 6). The public affairs, cultural, and news programs all provided extensive coverage of the new movements, but the call-in and free-form shows increasingly became a part of the counterculture itself. A second transformation was the way that the ad-libbing of these shows influenced many of the programmers throughout Pacifica, producing a more spontaneous and at times absurd tone compared to the austere intelligence that Hill had bequeathed the network. Over time, this informal, improvised, comic demeanor permeated the sound of FM music radio more generally, creating by the late sixties the genre known as "underground radio."

Alan Rich, KPFA's music director, 1954–1960. Rich developed innovative music programming and read children's stories such as *Winnie the Pooh* on the air.

Left to right: Vincent Price (actor and KPFA board member), Bill Butler (KPFA production director), and Harold Winkler (president of Pacifica) in the Pacifica Studios of KPFA for a special day of programming in September 1959. Photograph by William Caxton. Courtesy of Pacifica Radio.

Josephine Baker at KPFA, 1960. Courtesy of Pacifica Radio.

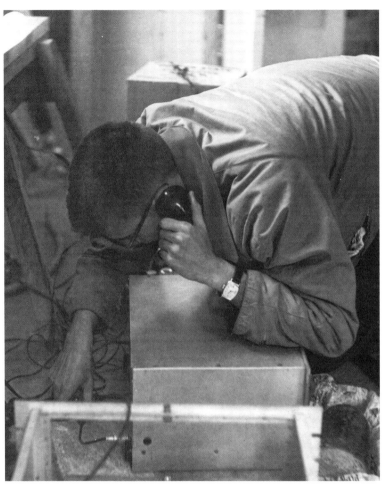

Watson Alberts volunteers at the opening of the new station, about 1951.
Courtesy of Pacifica Radio.

Left to right: Dr. Peter Odegard (KPFA Board), Frank Freeman (chair, Pacifica Board), and Lewis Hill in the mid-1950s. Courtesy of Pacifica Radio.

Chief engineer David Talcott makes adjustments on RCA 76B control board, about 1958. Photography by Marion Carnahan. Courtesy of Pacifica Radio.

Byron Bryan, early public affairs director, and guests in the studios in the 1950s.
Courtesy of Pacifica Radio.

Bill Triest, program director and founding member of KPFA, 1946–1953.
Courtesy of Pacifica Radio.

KPFA engineer Bruce Harris constructs the temporary transmitter building on Panoramic Way, about 1951. The tower fell over more than once. Courtesy of Pacifica Radio.

Lewis Kimball Hill, founder of Pacifica Foundation and KPFA-FM, in 1951.
Courtesy of Pacifica Radio.

Elsa Knight Thompson and unidentified discussion panel over KPFA air.
Photograph by Gerhard E. Gscheidle. Courtesy of Pacifica Radio.

> The social world may be uttered and constructed in different
> ways according to different principles of vision and division. . . .
> One may act by trying to transform the categories of perception
> and appreciation of the social world, the cognitive and evaluative
> structures through which it is constructed. These categories of
> perception, the schemata of classification, that is essentially the
> words, the names, which construct social reality as much as they
> express it, are the stake par excellence of political struggle.
>
> — Pierre Bourdieu, "Social Space and
> Symbolic Power"

A dynamic and radical version of the First Amendment stands at the heart of Pacifica's practices. Pacifica fused the Anglo-American libertarian creed of dissent as the lifeblood of democracy with a romantic notion of expression as the unique utterance of the soul. With roots in Emerson and Whitman, this aesthetic attitude has in fact shaped the majority of the programs in Pacifica's history. The experimental music, Beat poetry, and modern drama (often commissioned by the network and performed live in the studios) emphatically indicate Pacifica's devotion to free, creative, poetic speech.

One does not do justice to Pacifica's overall schedule by too neatly dividing between "politics" and "culture" (music, drama, poetry). Not only were many programs clearly syntheses, but the daily menu of political analysis and debate and cultural programming was complemented by more narrowly defined "educational broadcasts." Hearkening back to the ideals of the early broadcast educators, KPFA and later other Pacifica stations provided listeners with hours of "lectures" and entire courses (for example, the six-part series by David Reisman in 1958 entitled "Tocqueville and American Culture"). These topics could, at times, cover fairly arcane and complicated topics such as Dallas Smythe's ninety-minute lecture in 1962 on the geopolitical and technological significance of the Telstar satellite. To say that "only" 20,000 listeners may have tuned in to Smythe's program on WBAI (there is no way of really knowing — perhaps it was only 2,000) is not the correct way, or at least the only way, of considering the appeal of these programs. Even 2,000 is quite a sizable number to attend a "lecture." This listener-sponsored pedagogy exhibits neither dissent nor aesthetic expression; these programs demonstrate another facet of the First Amendment, often labeled the "free market of ideas." Pacifica at heart is the continuous weaving together of these different aspects of the free speech tradition throughout the years. Each broadcast day is a chorus of voices uttering the social world according to different principles: dissent, art, and *paideia.* This ongoing synthesis within Pacifica's programming— unprecedented in U.S. mass media—is the most complete measure of Hill's genius.

"THE SECRET OF LIBERTY"

The 1927 Radio Act that established the rules for government licensing of broadcasters explicitly stated,

> Nothing in this act shall be understood or construed to give the licensing authority the power of censorship over the radio communications or signals transmitted by any radio station, and no regulation or condition shall be promulgated or fixed by the licensing authority which shall interfere with the right of free speech by means of radio communication.[1]

Two factors—the prevailing sentiment that private, profit-oriented economic activity serves the commonweal and the ideology of the First Amendment—have kept government intervention into the media to a minimum.[2] According to critics such as Meiklejohn, this "freedom" has enabled commercial broadcasters to abrogate their public service responsibilities in their search for the widest possible audience. When asked recently about television's educational efforts, television producers pointed to such fare as the *Flintstones* and *Jetsons,* bringing to mind the fear of the earliest critics of commercialism:

> Our present radio setup which puts radio broadcasting in the hands of private radio monopolies deriving their revenue from advertising is dead set against the fundamental ideas which underlie modern civilization. ... For the first time in human history we have turned over the tender mind of a child to men who would make a profit from exploiting it—to men who have no real understanding of the consequences of their acts, for if they had, they would hang their heads in shame and make their apologies to the generations yet unborn.[3]

KPFA entered the airwaves at the point when alternatives to this situation appeared dim.

If the founding ideals of Pacifica hearkened first to Thoreau, Gandhi, and the great conscientious pacifist tradition, the earliest programmers were equally inspired by a fierce commitment to the Bill of Rights. Their "reading" followed from the public assembly rights demanded by early-twentieth-century radical movements such as the Wobblies, the pacifists, and other anarcho-socialist groups—a demand made often at their great peril. After World War I, a series of Supreme Court decisions acknowledged that the heroic civil disobedience and public agitation of these groups were emblematic of democracy itself, not a threat to the government but a realization of the fuller meaning of the Constitution. The jurisprudence from these decisions incorporated the Bill of Rights, most particularly the First Amendment, as a fundamental element of our constitutional order.[4] No one should be jailed for expressing radical political sentiments, a commonplace for agitators during the first decades of this century.

Justice Louis Brandeis cast democracy in the United States as an ongoing *performance* whose existence depended on dissidence and public dialogue.[5] In his interpretation of the "original intent" of the First Amendment, Brandeis wrote that our revolutionary leaders

> believed liberty to be the secret of happiness and courage to be the secret of liberty. They believed that freedom to think as you will and to speak as you think are means indispensable to the discovery and spread of political truth; . . . that the greatest menace to freedom is an inert people; that public discussion is a political duty; and that this should be a fundamental principle of American government. . . . Those who won our independence by revolution were not cowards. They did not fear political change. They did not exalt order at the cost of liberty.[6]

In the libertarian tradition from which Brandeis draws, the prophylactic ("Congress shall make no law") character of the First Amendment transforms into an affirmative injunction: "Public discussion is a political duty." The First Amendment's central meaning for democracy is the *obligation* for citizens to critically and collectively engage in the affairs of state. An "inert people" would lack the revolutionary will to transform unjust and antagonistic social relations.[7]

The vision of public discourse that spurs participants toward active citizenship and social amelioration vibrates at the core of Pacifica's approach to broadcasting. Over the past seventy years, there have been many arguments heralding the importance of the First Amendment, some cynical, others visionary, but few that match Brandeis's in so eloquently linking the act of speech with courage, democratic sovereignty, and revolutionary opportunities: not simply the argument that a free press enlarges the public sphere by providing citizens with information needed to govern, but the more radical understanding that the First Amendment both safeguards and implicitly promotes the sharing of opinion sympathetic to fundamental social reform. In the wake of radical social movements in the first decades of this century, the First Amendment was recognized as one key to unlocking "the rebellious instincts within us all."

An adherence to the dynamic virtues of free and robust speech can be found clearly in Hill's original conception of Pacifica. In an important statement on the affirmative quality of the First Amendment, he wrote that

> KPFA has set out to be an honest experiment in human understanding, a medium through which every experience, idea, or emotion may be conveyed without let or hindrance from its creator and interpreter to its ultimate receiver. . . . If a person has something interesting or important to say, it should be heard. But more than that, something should *happen* on KPFA's programs.[8]

In all likelihood, the most important influence on Pacifica's original First Amendment practices was Alexander Meiklejohn, the legal philosopher who helped write the initial bylaws of the foundation and remained a friend and adviser for the staff for years.[9] Meiklejohn, who lived in Berkeley during KPFA's embryonic period, was deeply distressed by the media's failed potential. In limiting the broadcast of political opinion, the commercial media had reneged on their promise to serve in the public interest. Under the guise of reaching and freely entertaining a mass public, broadcasters refused to live up to their civic responsibility.

As his jeremiad quoted earlier indicates, Meiklejohn retained a profound notion that radio could operate quite differently, and this only increased his agitation.

> The radio as we now have it is not cultivating qualities of taste, or reasoned judgment, of integrity, of loyalty, of mutual understanding upon which the enterprise of self-governing depends. On the contrary, it is a mighty force for breaking them down. It corrupts both our morals and our intelligence. And that catastrophe is significant for our inquiry, because it reveals how hollow may be the victories of the freedom of speech when our acceptance of the principles is merely formalistic. . . . We have used [the First Amendment] for the protection of private, possessive interests with which it has no concern. It is misinterpretations such as this which,

in our use of radio, are giving the name "freedoms" to the
most flagrant enslavements of the mind.[10]

Meiklejohn's writings are often called the "political speech" or "collective de-
liberation" approach to the First Amendment because he specifies a certain type
of expression — that which addresses affairs of state — as deserving greatest First
Amendment protection, more than "merely" personal expression. In his most famous statement, Meiklejohn argued that

> what is essential is not that everyone shall speak, but that
> everything worth saying shall be said. . . . no suggestion of
> policy shall be denied a hearing because it is on one side of
> the issue rather than another. The principle of the freedom
> of speech springs from the necessities of self government. . . .
> It is a deduction from the basic American agreement that
> public issues shall be decided by universal suffrage.[11]

When first enunciated in the late 1940s, Meiklejohn's was a bold response to the
fetters on public discussion emerging from the struggle against the spectral men-
ace of Communism. (See chapter 3.) With the onset of the Cold War, the government claimed that matters of international politics had become too complex for
the individual citizen, an argument swallowed almost in its entirety by the commercial media.

For a KPFA program on the Bill of Rights (the program won the Ohio State
Award for broadcasting excellence in 1956), Meiklejohn elaborated his theory
about speech and self-government. In this program, he argued that the overarching intent of the Constitution lies in subordinating the institutions of government
to the will of the people. Note how he, like Brandeis, ties the First Amendment to
its revolutionary origins.

> "We the people do hereby ordain and establish" are the rev-
> olutionary words which establish the freedom which is de-
> fined by the First Amendment. . . . The legal powers of the
> people of the United States are not granted to them by oth-
> ers. . . . All authority is ours alone. We are governed, directly
> or indirectly, only by ourselves.[12]

The First Amendment invites us to consider a utopian form of government: democratic sovereignty. "The primary purpose of the First Amendment is, then, that
all the citizens shall, so far as possible, understand the issues which bear upon
our common life. That is why *no idea, no opinion, no doubt, no belief, no coun-
terbelief, no relevant information, may be kept from them.*"[13]

This sensibility reverberates in one of Pacifica's earliest statements of purpose. For example, in 1946, Pacifica would strive

> to promote the full distribution of public information; to obtain access to sources of news not commonly brought together in the same medium; and to employ such varied sources in the public presentation of accurate, objective, comprehensive news on all matters vitally affecting the community.[14]

KPFA would broadcast

> genuinely significant choices... concerning the lasting understanding between nations and the study of political and economic problems.... It wished to treat proposed solutions primarily on the community and individual level.[15]

On the one hand, there is the smug assurance that Hill or his staff could determine and then broadcast those "genuinely significant choices," but on the other, one can also recognize the invitation for listeners to engage the vast geopolitical problems before them as the ultimate legislators.

Before continuing, one might note two critiques of this interpretation of the First Amendment. Some scholars believe that Meiklejohn's idealism is too narrowly aimed at affairs of state. Democracy, in his view, is not the complex set of cultural arrangements and personal opportunities invoked by Whitman, Dewey, and others, but essentially representative government, secured by the Constitution, enacted by the people. For this reason, speech that did not impact affairs of state seemed of less interest to him; it deserved protection, to be sure, but need not be guarded as scrupulously as other discourse that impacts policy making. (The discussion of the Carlin case at the end of this chapter demonstrates the pitfalls of this two-tiered evaluation of speech.)

A second, more significant criticism has been leveled at both Meiklejohn and Hill, one contrasting their unself-conscious elitism internal to their passionate democratic yearnings. If freedom of speech ensured that "everything worth saying" was given a fair hearing, "worthwhile to whom?" has been a question resonating throughout Pacifica's history. The next chapter discusses some of the contentious dilemmas this issue provoked. At all stations, and most dramatically in Berkeley and New York, the economic necessity of incorporating unpaid community labor for all levels of station activity, and the opening of unused evening and early-morning time slots to motivated volunteers, strongly modified Hill's original vision. Although Pacifica would continue to "employ such varied sources in the public presentation ... on all matters vitally affecting the community," over time, different groups and communities would come to nominate their own spokespeople, answering the question of "worth" in challenging, indeed contentious, ways. This process democratized and transformed the original paradigm in which radical autocrats such as Hill and Thompson served as the final judges of "worthwhile" programming. This transformation came at the expense, according to some, of balkanizing the stations. (See chapters 6 and 7.)

"The Rebellious Imperative of the Self"

During Pacifica's formative years, Meiklejohn's counsel was a major beacon guiding Pacifica's active dissent from the stifling Cold War rhetoric. The trail blazed by Madison and pursued by Brandeis and Meiklejohn reminds one of the radical pacifist potential of the First Amendment: its sponsoring of the peaceful, discursive formation of the general will. An uncensored free press, unrestricted public gatherings, free speech, and unregulated opportunity for pursuing spiritual paths provide citizens with the basis for popular sovereignty and representative government. The First Amendment in several dozen words codified the conditions for a democratic public sphere, a space of resolution of difference through dialogue rather than violence. The guarantee of security for principled, conscientious dissent — heralded by the Levelers, the Diggers, and other early modern protesters — prevents representative government from devolving into a tyranny of the majority. Pacifica surely took this Anglo-American libertarian tradition to heart and promoted it as widely as their signal could beam. However, the early programmers were not interested in politics alone.

Hill, McKinney, America Chiarito (the first music director), and the hundreds of performers and educators who sponsored the station and performed there held a vision of the First Amendment that transcended the more deliberative decision-making model invoked by Meiklejohn's writings. To what avail is the First Amendment's contribution to popular sovereignty and a government limited by checks and balances? Ultimately the "cash value" of the Bill of Rights is the opportunity it provides for self-assertion and self-reformation. From the start, Hill, with his commitment to anarcho-pacifism and his aspirations toward poetry, saw free speech on the airwaves as the opportunity for creativity — not simply adding one's voice to a democratic consensus, but the chance for unique human expression.

It is possible for a nation to possess a wide range of political liberties but still lack a culture that promotes the full play of human potential. Consider in this respect John Stuart Mill's warning:

> [Society] practices social tyranny more formidable than many
> forms of political oppression, since, though not usually up-
> held by such extreme penalties, it leaves fewer means of es-
> cape, penetrating more deeply into the details of life and en-
> slaving the soul itself.[16]

Constitutional guarantees that freed speech and press from censorship protected
the public sphere from undue state interference. But it would take more passion
and vigor — Whitman's "barbaric yawp" — to destroy the insidious, pervasive
tyranny of social norms.

In a KPFA program from the late fifties, Supreme Court Justice Hugo Black
expounded the "basic meaning" of the Bill of Rights in terms that drew from
Meiklejohn but also carried intimations of the radical yearning of Whitman. Black,
with a passion that loses its flavor and soft southern accent on the page, heralds
the virtues of the First Amendment. Providing a "plentiful degree of speculative
license,"

> Our First Amendment was a bold attempt to establish a coun-
> try with *no* legal restrictions of *any* kind on the subjects
> people could investigate, discuss, and deny. The framers
> knew, better than perhaps we do today, the risks they were
> taking in this.
>
> They knew that free speech might be the friend of change
> and revolution, but they also knew that it was always the
> deadliest enemy of tyranny. With this knowledge, they still
> believed that the ultimate happiness and security of a nation
> lies in its ability to explore, to change, to grow, and cease-
> lessly to adapt itself to new knowledge, born of inquiry,
> free from any form of control over the mind and spirit of
> man. . . . They were not afraid for men to be free. We should
> not be.[17]

Throughout the repressive environment of the fifties, an increasing number of
Pacifica programs strove to exhibit the dynamism ("explore," "adapt," "change,"
"grow") of Black's jurisprudence.

How might radio model an expressive, creative freedom muscular enough to
stand up not just to a repressive state, but also to the more invidious hegemony of
opinion and taste? The first step was listener sponsorship. Unhampered by the severe, if tacit, restriction placed on commercial programs, KPFA's music and literature shows, approximately 70 percent of its schedule, flew in the face of the formats and clichés of American broadcast entertainment. The programming, at its

best, celebrated eccentricity and the spontaneous exhibition of creative energy. Nearly daily, in-studio performances gave "KPFA complete uniqueness in this field, and made it the center of a continuous, area-wide music festival."[18]

Being open to the novel also meant taking risks with performers who had no established audience or pedigree. Throughout the fifties, broadcasts of the austere and formal experiments of T. S. Eliot, Schoenberg, and other modernists much admired by Hill gave way to the polymorphous cultural eruption of the Beats, where jazz, poetry, and sound effects filled the airwaves, producing an unprecedented aural environment. Not just political agitators but poets, musicians, performers, and other "friends of change and revolution" were treated with respect and given the opportunity to nurture their audiences on Pacifica. (That each flourished symbiotically in the Bay Area is due in no small measure to the fortunate geographical circumstances noted in chapter 3.)

The Beat performance aesthetic, based on an explosive improvisational assemblage of sound, rhythm, and jest, contained a distinct politics, one of the first countercultural attacks on the nuclear age.[19] Although rarely cast in direct political terminology, the Beats' overturning of received musical and literary modes was an implicit critique of both mass culture and nuclear exterminism. Norman Mailer's seminal 1958 essay "The White Negro" stands as one of the few explicit articulations of this cultural politics. As such, it bears some scrutiny. Mailer specifies the apocalyptic imagination within the flamboyancy of Beat culture:

> The American existentialist—the hipster, the man who knows that if our collective condition is to live with instant death by atomic war...or with a slow death by conformity with every creative and rebellious instinct stifled...then the only life-giving answer is to accept the terms of death, to live with death as immediate danger, to divorce oneself from society, to exist without roots, to set out on that uncharted journey into the rebellious imperative of the self.[20]

This urgency of keeping the horrific possibility of "instant death by atomic war" firmly in mind was made all the more difficult by the dominant response of consumerism and conformity—the stifling "social tyranny" that, as Mill warned, serves to deaden the soul itself. In this period, the many dissident thinkers and performers who struggled to break through "slow death by conformity" had almost no public airing on the mass media other than KPFA and later KPFK and WBAI.

Pacifica's increasing broadcasts of contemporary jazz and poetry, combined with public affairs and commentary, invited listeners to "set out on that uncharted journey." This may very well not be a journey we all wish to take. Nonetheless, it is essential that those willing to take the risk have the opportunity, for

> the eager and often inconsiderate appeals of reformers and revolutionists are indispensable to counterbalance the inertness and fossilism making so large a part of human institutions. The latter will always take care of themselves — the danger being that they rapidly tend to ossify us. The former is to be treated with indulgence, even respect. As circulation is to air, so is agitation and a plentiful degree of speculative license to political and moral sanity. Indirectly, but surely, goodness, virtue, law, (of the very best) follow freedom. These, to democracy, are what the keel is to the ship, or saltiness to the ocean.[21]

Not all the creative ferment of this period was cast in Mailer's apocalyptic terms, however. Much Beat poetry drew deeply from a Whitmannian "speculative license," all the more luminous in the face of impending doom. Consider, for example, this segment of Lawrence Ferlinghetti's poem "Great Chinese Dragon," broadcast in 1959:

> and he is a big red table the world will never tilt and he has big eyes everywhere thru which he sees all womankind milk-white and dove-breasted and he will eat their waterflowers for he is the cat with future feet wearing Keds and he eats cake out of pastry windows and is hungrier and more potent and more powerful and more omnivorous than the papier mâché lion run by two guys and he is the great earthworm of lucky life filled with flowing Chinese semen and he considers his own and our existence in its most profound sense as he comes and he has no Christian answer to the existential question even as he sees the spiritual everywhere translucent in the material world and he does not want to escape the responsibility of being a dragon[22]

In a trajectory from Whitman, through the Jazz Age, to the Beats, American performers and artists have struggled to see the world through "big eyes." Ferlinghetti's invocation of "the spiritual everywhere translucent in the material world" might be recognized as an inference toward the distinctive, as yet unrealized, potential of democratic personality, expressing poetically the utopian vision that underlay Pacifica's definition of broadcasting in "the public interest."

This particular broadcast is significant in Pacifica's history because the hour-long documentary on Ferlinghetti (with its "flowing Chinese semen") was cited in 1960 in the network's first major challenge from the FCC, who considered this poem, among others, "indecent." The commission had apparently received several letters from listeners objecting to the broadcast of Ferlinghetti's poetry reading, which included "Great Chinese Dragon" as well as "A Coney Island of the Mind" and "Tentative Description of a Dinner to Promote the Impeachment of President Eisenhower." These poems contained, according to the commission, "material that was vulgar, obscene, indecent, ribald, and in bad taste."[23] It turned out that the program under assault had previously been recorded by Chicago station WFMT; the unedited tape of that program was then played on KPFA.

In a rare contrite correspondence between the foundation and the commission, Pacifica responded to the FCC that

> there is much in literature and poetry which though not obscene or profane may still be offensive to many people.... Reexamination of the Ferlinghetti tape discloses that some passages in it do not quite measure up to Pacifica's own standards of good taste.... As events turned out, Pacifica was mistaken in not having carefully screened the tape.... In the future, Pacifica will not rely on other's standards, but will subject all programs to scrutiny to make sure they meet Pacifica standards.[24]

The commission ultimately dropped its charges, but that this poem could be judged as somehow inimical to the "public" should well cause one to ask, "Which public, whose interest?"

The lessons of KPFA's first decade of "free speech radio" are that neither constitutional protection nor the exuberance of poetic genius alone suffices to maintain the plasticity of intellect that democracy both invites and demands; only, as Dewey claimed, an explicit, collective effort "on as many fronts as culture has aspects" serves a truly democratic society.

FSM

Burton White, a student activist who later became KPFA development director in the mid-sixties, maintains that Berkeley's centrality for the entire decade's political and cultural explosion was due in no small measure to the "ambience KPFA created in the Bay Area."[25] White may not be precisely neutral, and *ambience* is not a scientific category; yet to others as well, it is hardly coincidental that Berkeley's free speech movement (FSM) emerged several blocks from KPFA's studios. It takes no great leap of imagination to consider the impact of the years of KPFA's programs stressing the importance of dissent on the students and their faculty supporters. Indeed, many Berkeley faculty had been regular commentators or guests on KPFA ever since the station had first opened.

The right to distribute literature in a public area in front of UC Berkeley's main plaza seemed an uncontroversial student demand. However, it rapidly escalated into a massive confrontation with the police, with the university, and with technocratic capitalism more generally. That it did so confirms the insight of Brandeis and others who noted in the First Amendment a more disruptive potential underneath its passive language.

Over the course of the sixties, Pacifica produced several shows about the FSM that highlighted the expansive and dramatic meanings the students gave to the category of "speech," in which civil rights, antiwar, and educational issues were all bound together. Betina Aptheker, narrator of an extensive 1966 program commemorating the FSM,[26] describes the series of events leading up to the actual protest: the dramatic anti-HUAC demonstrations in 1960 (broadcast on KPFA) closing the Red Scare in the Bay Area; the 1963 anti–Vietnam War protest demonstration—perhaps the first in the country—organized by the Berkeley [W. E. B.] Du Bois club when Madame Ngu received an honorary degree from the university; the struggle over integration in employment at a Bay Area hotel; the ongoing SNCC sit-ins; and the 1964 presidential elections. When, on 30 September 1964, the university prohibited the distribution of any literature from civil rights or antiwar groups on campus property, apparently at the behest of local business interests and members of the university board, students from a range of different social and political organizations united in an ad hoc "free speech committee" to protect their right to present their views. "The reason was very obvious. We weren't fighting over five feet of turf, we were trying to establish the principle of political organizing on campus."[27]

In the radio documentary, live coverage of the dramatic moment when students surrounded and took over a police car, beginning a siege that lasted for thirty hours, segues into speeches by university president Clark Kerr and FSM leader Mario Savio. Savio's blistering call for students to put their "bodies upon the gears and the wheels and the levers and all the apparatus of the machine" not only galvanized the Berkeley campus but set the tone for student rebellion for the next decade. Indeed, Savio's invocation transformed the issue. Neither the turf, nor the First Amendment, nor even the principle of political organizing was central:

at heart, the FSM was a struggle to shut down the "machine" of late capitalist knowledge industry, a prophetic movement that comprehended the increasingly central role of universities in economic and military affairs.

Much of the four-month protest itself found its way immediately onto KPFA's airwaves. The only media the demonstrators permitted to cover their meetings and planning sessions were reporters from Pacifica. During the takeover of the administration building in December 1964, KPFA ran a shuttle service for its reporters, who would rush their unedited tapes directly to the station, where they would be played within an hour. As Shana Alexander wrote in *Life* magazine:

> Among its other more obvious firsts, the Free Speech Movement at the University of California and its climactic revolt against the administration was probably the first full-scale revolution to be carried on radio in its entirety, and I found the tapes memorable listening.[28]

Alexander notes the "wildly mixed feelings of outrage and relief" that KPFA's broadcasts provoked: outrage over the bumbling administration, relief in the intelligence and idealism of the students, "a scruffy looking bunch, fond of flowing hair and beards, ponchos and army boots...a gathering of beatnik apostles."[29] This (nearly live) coverage of the FSM would be a harbinger of Pacifica's future reporting of the uprisings and demonstrations of the sixties. Reporters would continue to act much as they did in this instance—as participant-observers, chronicling events for broadcasts that would, when aired, serve as further prods to mobilization. A 1965 letter from foundation president Hallock Hoffman to the FCC in an application for license renewal further highlights the significance of the FSM broadcasts:

> In addition to covering the events, we broadcast the entire session of the [Berkeley] Academic Senate, for example,... as well as the meetings the President addressed, many student meetings, and many addresses of the student leaders. When CBS was putting together its documentary on the FSM, it came to KPFA for original tapes, and during the trials of the students, the court relied upon KPFA tapes to determine the facts of several matters.[30]

In her brilliant discussion of the FSM, Wini Breines provides an overview that could well serve as a description of the dialectic of free expression on Pacifica:

Reflecting on the Free Speech Movement, two meanings of "free speech" present themselves. "Free" in the civil liberties sense, so that citizens and students were able to express their political ideas and platforms without fear of punishment; and "free" in the sense of liberated, unrestricted communication that may foster political forms that transcend the existing framework, enabling utopian ideas to inspire individuals toward becoming political actors on the basis of their unmet collective and individual needs.[31]

In the next chapters, we shall return to Pacifica's free speech radio in the sixties and seventies, recast simply as "free radio" as New York's WBAI found itself pulled ever more into the vortex of the antiwar movement and the counterculture.

"THEY MUST *REALLY* BE BAD"

This chapter's last section returns to the issue of explicit government attacks on Pacifica. In 1971 a Senate committee questioning the FCC about modifications to the communications act delayed its overall discussion and took the commission to task for not censuring Pacifica for a program that broadcast the poem "Jehovah's Child," a work portraying Christ being fellated while on the cross. The classroom teaching of the poem in question had led to the dismissal of a professor at a local community college. This action in turn had occasioned a certain degree of public outcry. The program under attack was broadcast on KPFK at 10:30 P.M. It was a panel discussion on the firing of the professor and the nature of profanity, academic freedom, and the history of blasphemous language (a fascinating topic), during which two college professors and a clinical psychologist discussed the poem's merits. The poem itself had been read on the air only after frequent disclaimers and warnings about its subject matter.[32]

Nonetheless, this "abomination" overwhelmed at least two legislators, Senators Gurney and Pastore, during a 1971 Senate hearing on the state of broadcast regulation. In a sputtering hour-long denunciation captured on the tape "Obscenity and Pacifica," these legislators chastised the FCC for not reprimanding KPFK. Senator Pastore questioned how the FCC might justify granting Pacifica its Houston license when,

> as Billy Graham has warned, if television and radio can pose as an instrumentality, under the guise of art, to permeate salaciousness and obscenity throughout the fabric of our society, then a serious mistake will surely be made.

Senator Gurney was

> shocked to say the least that this sort of smut was aired over
> the radio waves, but even more shocked that the FCC granted
> a continuing license to an operation where programs of this
> sort are aired, which show the Lord in a perverted act....
> Why was this license (for the new Houston station) issued?

The FCC's majority response is exemplary in its support of the network. All seven commissioners were present at the hearing; only one commissioner recommended suspending Pacifica's license on the basis of criminal obscenity.[33] He explained to the Senate panel that the commission could not pursue this path after it had learned that the Department of Justice had determined that the poem's offensive language did not appeal to "prurient interests." Commissioner Cox, speaking for the majority, reminded the senators that there is a difference between obscenity, whose sexual imagery lacks any "redeeming social values," and indecency, far more loosely defined as that which "disregards community norms." Technically, the poem and program might indeed be "indecent," but they are not "obscene." Only "obscenity" is subject to legal action, something the senators obviously knew. In 1971, for courts to rule on obscene speech, the language must appeal to prurient interests in a manner patently offensive to community standards with no other redeeming merit.[34] This legal definition, and the fact that the criteria of "offensive language" are so vague, would become the central issues of the more widely known Carlin case two years later.

The FCC rightly called attention to the fact that the senators were attempting to ignore the de jure definition of obscenity and condemning the program in question based on their own religious and moral sensibilities. "This is simply a matter of taste, which we are not called upon to judge," Commissioner Cox explains. He lists the many merits of Pacifica's programs, its classical music and public affairs shows, which provide

> a range of service to the community that is all too often lack-
> ing on commercial stations.... In the nature of the service
> that Pacifica performs, it is bound to tread on the sensibili-
> ties of some people. Many commercial stations operate un-
> der the assumption that no one should be offended.

The legislators would have none of this.

> This is your typical "filthy word" station. But you're help-
> less unless the Attorney General gives you permission to
> prosecute? You're losing sight of common sense, *common
> sense*! We're all grown-up people here, and we know what

> obscenity is. We know what art is. And you have a station
> that puts on records with four-letter words, again and again
> and again and again.

One can hear the senator's wrath rising: "Oh my goodness, by the same token, they can show 'I Am Curious, Yellow' on television tomorrow! It frightens me. It frightens me to no end."

This passion against the "smut" of "Jehovah's Child" led the hearings to an overall attack on Pacifica, concluding with the suggestion that the FCC should perhaps show greater care following the trail of Pacifica's funding. In a long discussion concerning Pacifica's tax records, the senators intimated, like *Counterattack* a decade earlier, that there is something suspicious about listener sponsorship. Is Pacifica really an "educational" foundation? the senators queried. Raising the history of earlier complaints, including those attacking the Ferlinghetti poem, Pastore and Gurney fulminated, "You all seem perfectly helpless to act in the public interest, and that frightens me....To me, it's a question of guts....If a station is not living up to the law, then just refuse to renew it." That the station was by statute "living up to the law" of obscenity did not matter.

Toward the end of the hearing, Nicholas Johnson spoke for the majority of the commission when he claimed that letters about Pacifica comprised only a small fraction of the total number of complaints that the commission received. The programming of offensive subjects on Pacifica hardly equaled, in the commissioners' judgment, the truly enormous problems that they faced in 1971, such as escalating charges of racism in programs and commercials, government scrutiny of media violence, and public concern over children's programming. Although none of this assuaged the passionate concerns of the Senate panel, the discussion ended for the moment. In 1964 and again in 1971, the FCC demonstrated fairly strong resolve in upholding the right of Pacifica to broadcast as it deemed appropriate and to retain its licenses. This relatively protective situation would not hold for much longer, however, leading to one of the most well known battles between the FCC and a broadcast licensee, the case known in legal literature as "Seven Filthy Words."

At 2:00 P.M. on 30 October 1973, WBAI's Paul Gorman played a recording of a twelve-minute live monologue by comedian George Carlin, "The Seven Words You Can Never Say on Television," or "Filthy Words," as part of a special program on contemporary attitudes toward language. On the record, Carlin's audience raucously appreciates his discussion of the words

> you couldn't say on the public airwaves, the ones you *defi-
> nitely* wouldn't say, ever....The original seven words were

"shit," "piss," "fuck," "cunt," "cocksucker," "motherfucker,"
and "tits." Those are the ones that will curve your spine, grow
hair on your hands, and maybe even bring us, God help us,
peace without honor, and a bourbon.[35]

(A later monologue listed more than two hundred words, apparently inscribed on
a lengthy scroll that was unfurled as part of the performance.)

According to the Supreme Court notes,

> A father who heard the broadcast while driving with his
> young son complained to the FCC, which...issued a de-
> claratory order granting the complaint. While not imposing
> formal sanctions, the FCC stated that the order would be
> "associated with the station's license file, and in the event
> that subsequent complaints are received, the Commission will
> then decide whether it should utilize any of the available sanc-
> tions it has been granted by Congress."...The FCC charac-
> terized the monologue as "patently offensive," though not
> necessarily obscene, and expressed the opinion that it should
> be regulated by principles analogous to the law of nuisance
> where the "law generally speaks to channeling behavior
> rather than actually prohibiting it." The FCC found that cer-
> tain words in the monologue depicted sexual and excretory
> activities in a particularly offensive manner, noted that they
> were broadcast in the early afternoon "when children are
> undoubtedly in the audience" and concluded that the lan-
> guage as broadcast was indecent and prohibited.[36]

A lengthy legal battle ensued. The FCC's 1973 ruling, based on indecency
claims, was overturned in appellate court by a two-to-one margin, only to have
the Supreme Court, in 1978, back the FCC. Penned by Justices Stevens and Powell, the Court's five-to-four decision concurred that the FCC had legal jurisdiction to condemn the broadcast as indecent, and, more important, that in certain
instances the broadcasting of indecent speech did not deserve full First Amendment protection. The overriding significance of the Carlin decision for the legal
community lay in the Court's affirmation that the FCC has a broader scope in
prosecution of "indecent" language than is permitted outside of broadcasting. The
second repercussion concerned the peculiar manner by which the Court judged
the nuisance character of a broadcast. The majority ruling treated a program voluntarily received—the plaintiff had to turn his or her radio on before the offense

could occur—in the same vein as a range of involuntary violations of privacy (nuisances).

The First Amendment jurisprudence underlying the majority's reasoning in the Carlin opinion was influenced, ironically, by Meiklejohn's two-tiered evaluation of speech. This "sliding scale" presumes that the speech with the greatest social value is that which impacts political processes. Literature, scientific papers, and other "nonpolitical" writing is judged as less deserving of overall First Amendment protection. This line of reasoning, especially its application in the Carlin case, has been almost uniformly condemned by legal scholars as being "inconsistent with precedent and traditional First Amendment doctrine . . . and afford[ing] inadequate protection to First Amendment rights."[37]

The second half of the Carlin opinion upheld the FCC's actions by claiming that the "Filthy Words" program, broadcast during the daylight hours, would likely be heard by children and hence constituted an intrusion on privacy. Justice Powell argued that WBAI's actions were a "nuisance" for parents attempting to shield their children from vulgar language. In this opinion, the unwritten "right to privacy" was seen to override the First Amendment. This ancillary argument was necessary to gain a majority on the court.

The legal community's attack on the Pacifica opinion was summed up in a dissenting opinion by Justice Brennan, in which he called the ruling "disingenuous as to reality and wrong as a matter of law."[38] Picking apart the two pieces of the majority opinion, Brennan was especially caustic toward Powell's claim that "there are few, if any, thoughts that cannot be expressed by the use of less offensive language," intimating that Carlin's monologue would have the same quality and meaning if he had substituted polite terms for his "filthy" ones. This

> ethnocentric myopia enables the Court to approve the censorship of communication solely because of the words they contain. "A word is not a crystal, transparent and unchanged, it is the skin of a living thought and may vary greatly in color and content according to the circumstances and time in which it used." The words that the Court and the Commission find so unpalatable may be the stuff of everyday conversation in some, if not many of the innumerable subcultures that comprise this nation. . . . Today's decision will thus have its greatest impact on broadcasters desiring to reach, and listening audiences comprised of, persons who do not share the Court's view.[39]

Legal scholars have noted that fortunately the FCC "seems to have backed away from the precedent and been reluctant to apply Pacifica vigorously." Although the implicit chilling effect is impossible to gauge, "review of subsequent cases has shown that the great fear of censorship and the concern about First Amendment protection when Pacifica was first handed down were unwarranted."[40]

No commentary on the case has recognized that the FCC's attack on WBAI closely mirrored the position that the senators debating SB 2004 took in 1971. During the earlier debate, the senators all but demanded that the FCC use its authority to attack indecency on the airwaves, regardless of the legal precedent. And indeed the majority Court decision in the Carlin case chose to follow precisely this path; when Justice Powell claimed that the "offensiveness" of Carlin's monologue was equivalent to that of obscenity, regardless of the broadcast's potential social merit, he was speaking for the senators. In his highly tendentious conclusion, Justice Powell argued that Carlin's "Filthy Words" could play "no essential part of any exposition of ideas, and are of such slight social value as a step to truth, that any benefit derived from them is clearly outweighed by the social interest in order and morality."[41]

This "social interest in order and morality," a term adapted from an earlier case, was precisely the motivation behind the senators' 1971 attack on Pacifica's "smut," when Senator Gurney worried so greatly about the "rot permeat[ing] through the fabric of society which will destroy society." As that hearing so clearly indicated, the rot he had in mind was the sort of attack on propriety exemplified by Carlin's speech. There is no explicit evidence that the FCC was moved by the legislators' arguments in 1971; indeed, at the time, they stood firmly in the network's defense. Nonetheless, the commissioners took action against Pacifica for the second time in a decade shortly after an attack on the network by a Senate committee.

The Carlin case remains important and fascinating, notwithstanding the fact that it has yet to exercise the chilling effect predicted at the time. It is the only event in Pacifica's complex history that remains of ongoing scholarly concern because of the fear of some legal observers that the Court may at some point return to the highly suspicious language of a "sliding scale" of merit when determining the protection of speech in the new telecommunications arena, such as speech on the Internet, using the Pacifica decision as a precedent.

However captivating it may be, stressing this case risks focusing Pacifica's practices and Carlin's performative brilliance too narrowly within a legal idiom. As Schiffrin (and this chapter) has argued, "the First Amendment's cultural significance may be more important than its legal significance.... The social and political force of the First Amendment goes beyond its legal force."[42] The real issue behind the "Filthy Words" case is not the (questionable) legal reasoning of the Court majority but the relationship of speech to social taboos. Precisely by the furor that his monologue provoked, Carlin highlighted the overlapping elements of politics, psychology, religion, and ethics that constitute the matrix of speech.

Speech is situated in the center of a paradox. We exist, precariously as it were, between the need for uniqueness and the need for affiliation. "The self is individual and it is social. But the requirements of individuality are in conflict with the demands of sociability in a way that does not seem immediately capable of solution."[43] To strive for uniqueness minimizes our membership in community. Yet to conform to social convention in the desire for relationship diminishes the peculiarities of character that distinguish us as individuals. One of the central themes of this book is that Pacifica, by championing "free" speech and "really bad" language, enables an investigation of these issues while not resolving them. On the one hand, it enthusiastically encourages personal expression, fulfilling the "requirements of individuality." This is unquestionably what Hill had in mind when he claimed that listener sponsorship would allow Pacifica's announcers to express what they believed "to be real, good, beautiful and so forth and what they believe is at stake in the assertion of such values."[44] Yet the "stake" in any personal assertion of the "real, good, beautiful and so forth" can only be assessed collectively. The next chapter examines the genesis of community radio in light of this situation, detailing the struggle within Pacifica as the network's guiding vision shifted from its original use of the microphone to express individuality to a more consistent hearkening to the claims of collective sensibilities in which the self is embedded.

The scandal of Carlin's "filthy words" lies in their penetration through the veneer of custom to a primal moment when inappropriately uttered profanity or blasphemy could provoke divine wrath and render community inoperable. The terror that taboo language once provoked has over millennia "virtually disappeared, so that bad language ... is now at most only shockingly rude."[45] Nonetheless, in its profane mode, language reminds us of its primordial power, for it alone possesses the means whereby humans become persons and communities produce their webs of tradition and value. These phenomena ("wonders by the side of which transubstantiation pales")[46] are the implicit point in both the language of Carlin's WBAI skit and the legal judgments its broadcast occasioned. At the moment of performance, Carlin individuated himself precisely by daring to utter the forbidden. The Court spoke for the social, convicting him of defying the conventions of propriety and obedience that make collective existence, and the appraisal of value, possible. Robert Post, a scholar of the First Amendment, noted that

> we face the constant choice whether to design [First Amendment] doctrines to sustain the common, socially embedded identities of citizens, or instead to design them so to protect the space for autonomous citizens independently to create their own social arrangements.[47]

The legal system's debate over issues of obscene, indecent, and profane language, and the unique judicial arguments raised in the case of *FCC v. Pacifica,* might be understood as the particular point where this conundrum comes fully to bear.

6. WBAI and the Explosion of Live Radio

> The genius of the heart from whose touch everyone goes away
> richer... richer in oneself, newer than ever before, broken
> open, blown upon and sounded out by a thawing wind, more
> uncertain, perhaps, more delicate, more fragile, but full of
> hopes that as yet have no name.
>
> —FRIEDRICH NIETZSCHE, *BEYOND GOOD AND EVIL*

By the early sixties, the three stations in the Pacifica network had a coherent, if eclectic, schedule: music, poetry, and drama, lectures and discussions, and a wide array of cultural and political commentary. This challenging aural environment earned the network abiding loyalty from small, dedicated audiences in Los Angeles, the Bay Area, and New York. Over time, the civil rights movement, the war in Vietnam, and the upsurge of radical protest globally had immense impact on the network, altering its soundscape and ultimately giving birth to the genre known as "community radio."

Abstractly, the relationship between Pacifica's listener-sponsored broadcasting and its reputation as a founder of community radio need not be a complex issue. The stations in the Pacifica network have been supported, in the main, by committed listeners who live nearby, many of whom participated directly in station operations as board members, programmers, and volunteers. Since the early fifties, Pacifica listener groups would meet for informal discussion and fund-raising, a process repeating itself with each new station, fostering a sense of solidarity and community within Pacifica's audience. Guided by Hill's original vision, Pacifica's programmers sought to fulfill its mission to promote peaceful internationalism by first demonstrating a sense of responsibility to local listeners — acting out the familiar "think globally, act locally" idea long before the phrase became popular. A 1951 summary announced that KPFA had "summoned out of the community an enormous and varied energy, talent, goodwill and trust."[1]

However, this perspective on the roots of community radio is complicated by Vera Hopkins. The network's most thorough archivist of written memos and articles, having worked at KPFA for almost thirty years from the late 1950s until 1987, Hopkins has served as the institutional memory of Pacifica. In a 1983 letter to Larry Bensky, former station manager at KPFA, Hopkins emphatically argued that "in my bones I think of the 'community radio' aspect of KPFA as counter to what early KPFA sounded like and what the staff thought was the purpose of KPFA."[2] "Community radio" for Hopkins was signaled by the (disastrous) emergence in the late sixties and early seventies of "Third World" and "women's" departments at the stations, a restructuring that she dramatically calls *"non-Pacifica, . . . in contrast to departments by intellectual division: public affairs, Music; Drama and Literature, News, Children's Programs."*[3]

Hopkins is correct, to a point; when they pioneered listener sponsorship and politically committed reportage, Hill, Thompson, McKinney, and others imagined their "community" as one defined by intellect—the "educated minority"—rather than one defined by gender, ethnicity, race, or, arguably, even location. Yet although perhaps not entirely congruent with the original goals, neither does the unabashed pluralism emerging in the late sixties seem quite as divergent from the founding ideals as Hopkins claims.

Historically, the ideal of community and the abhorrence of violence and war were fused in the origins of Christianity. Paul's notions of "membership" derived from a ceremony of "communion." Believers were unified within the body of Christ into a community marked most emphatically by a refusal to bear arms. The key to Christian pacifism was a new bond of affiliation possible once the convert rejected the idolatry of warfare.

More recently, Dewey recognized that there was a deep affinity between the frailty of communal solidarity in modern life and the attraction of war:

> The most militarist of nations secures the loyalty of its subjects not by physical force but through the power of ideas and emotions. . . . The balked demands for genuine cooperation and reciprocal solidarity in daily life finds [*sic*] an outlet in nationalistic sentiment. . . . If the simple duties of peace do not establish a common life, the emotions are mobilized in the service of a war that will supply its temporary stimulation.[4]

Although these brief examples do not indicate the precise form "community radio" would take, they anticipate that a pacifist institution might well expend its effort to build and sustain a phenomenon called "community." Pacifica *is* lauded by scholars and activists around the globe for forging community radio, a rare form of media in a universe dominated by commercial and state control.[5]

This chapter will chart WBAI's effort to accommodate the bewildering prolif-eration of groups demanding a place in the schedule in the name of representing "their" community, a challenging and unresolved process. Was righteous indig-nation enough to guarantee that a producer was capable of handling the nuances of live broadcasts to hundreds of thousands of listeners? On what basis could a program director determine who could speak for a given group? In these circum-stances, for example, how was a male station manager in 1973 able to decide which of the two or three radical feminist groups demanding airtime spoke with greatest legitimacy about lesbian identity?

Joan Scott, in a recent discussion of multicultural politics, highlights the un-derlying dilemma Pacifica faced in the effort to accommodate new programmers:

> The fact of belonging to an identity group is taken as author-ity enough for one's speech; the direct experience of a group or culture . . . becomes the only true test of knowledge. The exclusionary implications of this are twofold; all those not of the group are denied even intellectual access to it, and those within the group whose experiences or interpretations do not conform to the established terms of identity must either suppress their views or drop out.[6]

Scott's succinct description of these issues is a useful introduction to the trans-formations that Pacifica and the New Left generally underwent in the decades from the sixties to the eighties. The manner in which these changes played them-selves out at Pacifica had a unique inflection, however. During this period, the bitter contention over ideology and the problems of "authenticity" were not sim-ply internal network affairs; the struggles over access to the microphone were of-ten broadcast over the air in raging polemics open for all to hear.

THE CABAL

Throughout the 1960s, WBAI played a central, unifying role in New York's vast counterculture. According to Larry Josephson, morning host and later station man-ager in this period, WBAI "helped make the sixties what they were in New York. Everybody listened."[7] Although precise audience figures are unreliable, in the late sixties, perhaps 600,000 tuned in to 99.5 each week for "free radio": news, agita-tion, music, and conversation, combined with live coverage of rallies, sit-ins, be-ins, happenings, protests, and street theater. This vast, devoted community of lis-teners was far and away the largest listener base of any Pacifica station (and one never duplicated subsequently). Many actively participated in all aspects of the station: as volunteer producers, calling in daily to the live shows, and in the events and demonstrations the station promoted. By 1971 almost thirty thousand chose to sustain the station by paying ten to thirty dollars annually as subscribers, enabling the station to expand its studios and modernize its archaic technical infrastructure.

An understanding of the constantly expanding and evolving role WBAI played in New York in the sixties begins with Bob Fass, the host of the program *Radio Unnameable,* now in its forth decade.[8] In WBAI program guides from the early sixties, Fass invited his audience to tune in (at 1:00 A.M.) to "listen to the grass grow."[9] What they heard was an unprecedented ("unnameable") melange of music, poetry, political analysis, interviews, stream of consciousness monologues, and phone calls from the audience—"the cabal." ("It comes from the word 'horse.' Originally people who rode out at night with their identity concealed—even from each other—to plot or plan something subversive. And I thought, 'That's it!' ")[10]

With his instinctive psychological acumen, his avuncular demeanor, and extraordinary dexterity on the soundboard, Fass was a master radio artisan. More important than his skill using WBAI's archaic equipment was the striking, ineffable immediacy of his show. Fass's deep baritone, coached by years of theatrical training, could both soothe and agitate simultaneously: it was an instrument well suited to mediate the utopian premises and political strategizing of the emerging counterculture.

The show had no format. Before Fass, WBAI had signed off around midnight every evening. When he volunteered to fill the early-morning hours in 1962, he was allowed to do whatever he pleased. No one would be listening, under any circumstances. Within two years, *Radio Unnameable*'s nightly extravaganza had emerged as a source of vital energy and imagination, where

> spontaneous and collective political actions unfolded, new modes of communicating, decision-making, and analysis developed, as did solidarity and new kinds of caring, which in turn became the basis of future protest and changing notions of politics.[11]

(In 1967 Fass's audience was large enough that the station hired a personal secretary to keep up with the amount of mail and phone calls his show generated.)

Fass's friend Bob Dylan was a regular guest, answering calls and chatting with his fans for hours; it is rumored that "Blowin' in the Wind" had its public radio debut on the show. Arlo Guthrie first sang his saga of "Alice's Restaurant" one evening in 1967; Abbie Hoffman, with whom Fass helped found the Yippies, called in daily during recesses from the Chicago Eight trial to give live updates of the court progress after Fass began to augment his nightly broadcasts with a daytime program as well. For the New York alternative scene, *Radio Unnameable* had become "more than a radio program: for the true believer it was a way of life, and Fass's devoted listeners followed him through his own changes in life style and politics."[12]

During the same period, WBAI provided the most extensive war and antiwar coverage in the metropolitan area in spite of the station's meager budget. In 1965 news director Chris Koch was one of the first Americans to produce programs directly from Hanoi, a clandestine, illegal feat for which he was fired from the station, only to return with honor after hundreds of subscribers rallied to his support. In 1967 a magazine reporter noted that the station owned "only four tape recorders, and half were in Vietnam."[13] Koch's trip to Hanoi had blazed a trail for other young WBAI reporters seeking to present a wider angle on the war. Dale Minor won an Armstrong Award for his on-the-scene coverage of the campaign around Da Nang in 1967. Seymour Hersh first broke the My Lai story on WBAI in 1969.

These famous highlights of WBAI's war reporting deserve the accolades they have garnered; however, they should not obscure a related effort at public service: the daily extensive coverage of multiple aspects of the war that all Pacifica stations provided. As a network, Pacifica was the only media outlet to make extensive use of Agence Française, the sole Western news agency with a permanent staff in Hanoi. WBAI's lengthy morning and evening news reports of the war were supplemented in 1967 by a new Washington news bureau, as well as by hundreds of special public affairs programs on government war policy. Combined with the station's concrete involvement with the burgeoning antiwar movement — live coverage of demonstrations, teach-ins, and strikes, and the extensive daily bulletin board and calendar service — the programming around the Vietnam War helped to build a huge listenership across the political spectrum who habitually listened to WBAI.

By 1967 a new generation of programmers — Larry Josephson, Steve Post, and, somewhat later, Black Power activist Julius Lester — all began experimenting with "live radio," a free-form style derived from Fass's brilliant improvisational work. Sometimes they would take phone calls or read the daily paper, mouth filled with danish; other times they would launch into lengthy, scathing commentary on government policy or the price of recreational drugs, or play the Beatles' "Lady Madonna" nonstop for several hours. While the experimental poetry and august forums such as "Television in Africa" retained their slots in the broadcast day, the live shows with Post, Josephson, Fass, and Lester, combined with the extensive war coverage, served to elevate the audience's expectations of the station. Before there was something specifically designated "community radio," WBAI's "live" or "free radio" programs, reaching out to a "cabal" of coconspirators, prefigured what such a thing could be.

Without fear of alienating advertisers, WBAI was uniquely situated to catalyze an immense, ongoing conversation among its audience, inviting them to share in their dreams and desires; in the midst of the inchoate energies that the sixties had stirred up, the station had assumed the immense responsibility of mediating the movement's understanding of itself in the New York megalopolis, nurturing innumerable "hopes that as yet have no name."

As its audience grew and the tenor of the times became more volatile, Fass led the station to a more active and public role. The first of the major spectacles Fass choreographed with *Radio Unnameable*'s audience was a "fly-in" at Kennedy Airport, an event setting the stage for New York's first "be-in" in Central Park a few months later. Fass spent the weeks of early 1967 inviting the "cabal" to venture to Kennedy Airport's international terminal on 11 February at 1:00 A.M. to admire the Calder mobile and the lights of the planes. An estimated three thousand participated. Recalls one participant:

> I went there on my own without knowing if anyone else would come. I walked into the terminal and sure enough, no one was there. But then I heard this applause, so I looked up. And all around the balcony were hundreds of people — freaks — greeting everyone who came with this ovation.... Everybody was high, everybody knew everyone else was high. Even though we had never met, because we were all connected through Fass, we felt connected to each other. It was like being part of the greatest party you could imagine, in the middle of this huge airport at one a.m.[14]

Steve Post, who programmed the early-morning slot on weekends, responded by calling for a Central Park "fat-in" several months later in which revelers burned life-size posters of Twiggy while feasting on ice cream. (To be sure, older, more sober-minded radicals, such as Dorothy Healey, found something askew with this entire situation. A humane Marxist agitator and organizer, with more than thirty years' experience hosting programs at various Pacifica stations, Healey would comment in 1993 that the whole New York scene in the sixties had an "obsessive childlike silliness" about it, an observation not meant as a compliment.)

These theatrical, participatory events originating from the station complemented other programming: the live round-the-clock coverage of the Columbia
student rebellion in 1968; the use of the station switchboard as an extensive calendar and hot line; and, somewhat later, the concerts, readings, and consciousness-
raising sessions held in the station studios before live audiences. Many thousands
of people

> always listened to WBAI for their politics . . . for information
> about community events, when the march was gonna be, when
> the rally was gonna be. . . . Hell, the whole Chicago Dem-
> ocratic Convention [protest] was organized by WBAI. . . . We
> all should have gone to Chicago and have been tried as con-
> spirators instead of just the Chicago Seven . . . *because we*
> *were all in on it.*[15]

Given Fass's participation in the founding of the Yippies, his close and abiding
friendship with Paul Krassner, Jerry Rubin, and Abbie Hoffman, and his personal
participation in the demonstrations in Chicago, which he was "covering" for the
station, this listener's exuberance may be only a slight exaggeration.

Unlike the educated elite that Hill saw his Berkeley station serving in 1950,
during an ongoing political witch-hunt at a time when less than 15 percent of the
audience owned an FM tuner, WBAI was a fifty-thousand-watt beacon in the
middle of the dial in the world's largest media market, broadcasting to an audience poised for revolutionary change.

SHARED INTIMACY

By the turn of the decade, most of the counterculture (and straight) community in
New York knew that WBAI's extensive call-in programs would provide a safe
space for someone to chat about being busted, their homosexual lover, or the problems with their rabbi with thousand of others listening and calling in to comment.
In the late sixties, new producers, notably gay and feminist activists, began blending their programs into a daily schedule with the older "personality-based" broadcasting, as critics would designate the shows of Fass and his epigones. The allure
of these newer shows rested more on the frank and often eloquent conversations
between the host and audience than on "free radio's" hallucinogenic blending of
music, sound effects, agitation, and spoken word. "Free radio" in the sixties invited listeners to become part of a vast underground, psychedelic conspiracy everyone sensed emerging all around; the later shows tended to be less flamboyant as

previously silenced, invisible, or ostracized groups emerged in the media lime-light and found in radio a unique and wonderful tool to identify themselves and to recognize and speak with each other. In 1970 it might have been difficult to make hard-and-fast distinctions between the different programs, mingled as they were during any given broadcast day, but over time the differences in both style and politics would widen.

As the sixties ended, these newer programs reached out to different listeners, expanding the distinct audiences the station served. It is at this moment, when Puerto Rican and black nationalists, Native Americans, radical lesbians, Asian-American activists, feminist spokespeople, and newly mobilized ecologists all began regularly scheduled programs that we find the earliest references to some-thing specifically designated "community radio." No single show or personality dominated the schedule; each individual announcer gained confidence from the sequence of so many different programs, one often following another. Callers and hosts engaged in an unrelenting, collective effort to speak honestly about simple issues such as housework or the negative stereotypes of Greco-Americans. As several now recall: "We felt we had no restrictions talking about values, trying to make sense out of our personal experiences."[16] "It's hard to imagine [as a woman] how different it was to hear someone talking honestly — about anything — on the air."[17] "You have some of these same gay shows today, but none of the feeling of things *crystallizing*. Back then every show was an experiment."[18] This final state-ment bears some attention. From the perspective of the present, after years of Oprah and Geraldo, certain of these programs may hardly seem as politically combustible as the exposé of the FBI's surveillance tactics or as magically sur-real as the "fly-in"; however, for the people who heard and responded to them in their genesis, and who ended up relying on them for nurture and education, they were of supreme significance in building identity and sustaining community.

By 1971 the schedule had expanded to accommodate as many as three or four slots a day for programs from the new social movements, but the inclusion of these shows was neither simple nor uncontroversial. Painful decisions were made determining which host or which show was most appropriate. Carolyn Goodman, an influential local board member, recalls:

> Pacifica throughout its history had a way of bringing on many voices. But voices changed somewhat in the late sixties, in the sense that it wasn't just a matter of expressing an opin-ion. They became in some instances angry voices and voices of communities *demanding* representation which they may not have had before.[19]

Critics have claimed some of the newer programs were hosted by people who knew little about "good radio." One listener wrote in 1972:

> WBAI has been...more fatuous in its thinking and plan-
> ning than I would have thought possible two or three years
> ago....the station, in an unthinking and somewhat lazy way,
> has become the special preserve of small coteries and little
> in-groups whose idea of radicalism is speaking to the con-
> verted....Your concentration upon matters that are periph-
> eral and even silly enervates and dilutes the attention that
> should be given to subjects that are desperately important.[20]

For the examples of silly and peripheral programs, this listener cites the prepon-
derance of "feminist orgasm worship" and his displeasure with Charles Pitts, the
host of the country's first regularly scheduled openly gay program. The writer
calls Pitts, an extremely controversial figure even for those who supported his
program, "spiteful, intolerant, and tedious, and a querulous spoiled brat to boot."

The station had been relatively solvent during the late sixties; it had launched
a successful major fund drive and in early 1971 moved its operations to a large
church in mid-Manhattan, a comfortable, well-equipped facility fondly remembered by all who worked there. The nearly thirty thousand subscribers who pledged
an average of fifteen dollars of yearly support in the early seventies was a vast
number—subscriptions plummeted to eight thousand in 1978 after the strike. The
generosity of some wealthy donors and extensive listener support provided the financial basis for WBAI's experiment to continue, albeit with ups and downs, at
least until 1974. However, several incipient difficulties then began to converge.

A foreshadowing of the problems to come occurred in 1969. A guest on Julius
Lester's program *The Great Proletariat Cultural Revolution* read an anti-Semitic
poem written by a black teen over the air. The poem had been a response to the
teacher strike and the turmoil in the Ocean Hill–Brownsville neighborhood over
community control of schools. Tensions had boiled over in a confrontation between the largely Jewish teachers' union and the local black community's desire
to hire and fire school personnel.

Lester, a prominent spokesman for the civil rights and black nationalist movement, was not himself advocating the vitriolic, pro-Hitler sentiments of the poem.
Nonetheless, the poem's broadcast elicited enormous response, leading to pickets
at the station and public denunciation of WBAI. The station took a principled
First Amendment stand, admitting that some of the audience was offended but arguing that the incident had been blown out of proportion by Albert Shanker and

racists within the teachers' union. In this imbroglio, which included death threats and other hints of violence, Lester's staunchest allies were the station's Jewish programmers—Post, Josephson, and others—while his comrades in the black community were almost altogether silent. (Lester subsequently converted to Judaism.) Although the station's mass subscription base did not at the time greatly suffer from this controversy, a core group of "wealthy Jewish communists" who had supported the station in the past became reluctant to give as much support.[21]

Margot Adler, feminist host of the pioneering early-morning call-in show *Hour of the Wolf,* had no doubt when the station fortunes began their decline. "The Vietnam War ended, and we lost half our audience. It was as simple as that. WBAI grew from the blood of the Vietnamese."[22] From Adler's perspective, the large audiences for both the "free radio" of the sixties and the newer "community" shows in the early seventies were for the most part drawn from listeners of the two daily news shows and late-evening war summaries. Few would dispute the assessment that much of the station's financial and popular support was, in the main, the fruit of WBAI's award-winning news team, which since 1965 had provided the area's most comprehensive coverage of both the war and the antiwar movements. This was the bread and butter of the station, the mass base from which the experiments in "free radio" could draw an audience. As Josephson put it:

> *Everyone* was against the war by 1970. All these people from
> Westchester and Great Neck didn't just listen, they subscribed
> to WBAI. This wasn't your hard-core New Left or counter-
> culture types from the Village [a reference to Fass's audience],
> but thousands of middle-class lawyers and teachers who liked
> to get high. They were the ones giving us all our money.[23]

With the fracturing of the larger antiwar movement in the early seventies, combined with the withdrawal of American troops from Vietnam in 1973, this audience began to dwindle.

Yet even a tiny audience by commercial standards might mean thirty thousand to fifty thousand listeners, a considerable number by the measure of the underground press or public rally. Precisely because the times were confusing—the mythic revolutionary moment was evaporating—both the producers at WBAI and their audiences depended ever more on the common etheric space in which they had habitually gathered to make sense collectively of what was happening.

Given Pacifica's overarching history of financial insecurity, WBAI's success in attracting subscribers and funding in the late sixties was a striking, fortuitous anomaly. As the red ink began to mount in late 1973, eliminating any of the paid staff was unthinkable. When confronted by the escalating financial difficulties in 1974, many salaried personnel opted for pay cuts. Several now claim that they would have been willing to work for nothing, believing the thrill of being on the air was compensation enough ("the greatest high of all"). Indeed, some key staff members did work for almost nothing — less than $10,000 per year for full-time work in 1975. For all this altruism, one must consider the assessment offered by Larry Josephson, the station manager in 1974, when the troubles began to threaten overall operations: "What about the secretaries and janitors? They're not on the air and they want to get paid every week. That's what everybody seemed to forget."[24]

Thus by the mid-seventies two phenomena intersected. For more than a decade, WBAI had experimented in fulfilling "the social destiny of radio" in an unprecedented manner. If not every experiment succeeded, this did not diminish the staff's collective sense of their accomplishment creating a vastly expanded, participatory public sphere. However, with this success came little wisdom about keeping the station viable when the audience and subscriptions waned.

As the belt tightening began, a series of charges of sexism, racism, and elitism aggravated an already charged environment. Although each charge had certain merit, the tenor of the debate was warped by the highly overwrought polemics of the time. It is doubtful that any of these problems alone would have pushed the station over the edge, but the combination proved devastating to morale. While much attention and programming addressed the issue of sexism and hetrosexism, it was the failure to deal with the issue of race that ultimately led to a crisis. The problem of elitism, cast in neo-Maoist terminology as an attack on "experts," was a permutation of questions that had been with Pacifica since its founding by mandarin radicals in the late forties.

By 1970 no one would dispute that WBAI for its first ten years had been dominated by a largely straight white male staff. Women programmers consistently faced a struggle to broadcast as equals in this heavily male environment. With the rising of the woman's movement, the situation began to change. In 1970 the most well-known radical-feminist broadcaster, Nanette Rainone, began producing *CR*, a show in which female listeners participated in a weekly on-air consciousness-raising session for two hours. Her program was simply "a group of women honestly discussing their lives."[25] In one of her more dramatic programs, women were invited to come with their mothers to the station studios for a gynecological self-help session. Unlike the late-night and early-morning call-in shows, *CR* was broadcast in the middle of the day, probably "touching and transform[ing] more lives than all of WBAI's broadcasting of the previous decade."[26] In a practice that caused much dissension, her calls were screened to prevent men from participating.

Rainone, who became station program director in late 1971, served as a model for a range of producers, men and women, as they filled the air with call-in shows and public affairs programs attempting to realize the feminist injunction to connect the personal and the political. Responding to charges that Rainone and other new programmers were diluting WBAI's "revolutionary" message, general manager Ed Goodman wrote in 1973 that

> people who were involved in 60's politics are now committed to a much less visible kind of personal politics and growth. It remains to be seen how important, in terms of impact on the whole society, this new manifestation of energy will be. However, one thing is clear. It is more difficult to translate these new developments into arresting, provocative, communicative radio. . . . It's not all out there in the streets anymore to cover.[27]

In trying to give voice to this more subtle kind of "personal politics and growth," WBAI was modeling a new form of inclusiveness in broadcasting, known later as "community radio."

Others found less merit in these ventures and were unwilling to defend Rainone, however, citing her tenure as program director as the moment when the station began to fall apart, with each host staking a claim on airtime without regard to overall scheduling blend. Whether the effort to bring feminist consciousness to WBAI was fondly or bitterly recalled, there is little dispute that both WBAI's programming and internal staff relations confronted and struggled with some of the core issues of feminism and patriarchy during this period, both on the air and within station operations.

The charges of elitism at WBAI were less a direct attack on Pacifica's "highbrow" approach to programming — which was far less hegemonic in New York in 1970 than it was in Berkeley or Los Angeles — than a reflection of the complex class dynamics, and the peculiar psychology of petit bourgeois guilt, found in the New Left more generally (and epitomized in the New York Jewish Left in particular). Most of the well-known WBAI programmers in this period were moderately affluent children of the middle class. Their political sensibilities were forged in the cauldron of the sixties, when participation in a range of civil rights and anti-imperialist movements led them to idealize bonds of solidarity with those outside their own class background. However well-intentioned and motivated they were, the staff collectively did not discover a populist, multicultural form of programming that might invite "the silent majority" to keep their radios tuned to

99.5 and participate in the political transformations toward which the programs hearkened. In this regard, it bears repeating that Hill, and earlier leaders of Pacifica, had no illusions that Pacifica's dissident and avant-garde programming would appeal to a mass audience.

The issue of elitism in New York had as much to do with internal station politics as with the caliber and intellectual complexity of the programs per se. The situation in the early seventies was summed up neatly by Josephson. In his analysis, the network had chronically suffered from a split between two groups: the "radio people and the politicos." Pacifica was founded and sustained by politically motivated media professionals — radio people, for whom basic broadcast standards mattered. Meeting technical and aesthetic norms did not contravene their political vision. Indeed, their genius was synthesizing the two. However, Pacifica, relying on volunteer labor and hence lacking normal criteria for screening employees, attracted many enthusiastic programmers who viewed "Pacifica" less as a radio network than as a bully pulpit for their various causes. This was what gave the network its dynamism and distinguished Pacifica from any other mass media in the United States. It also accounts for the network's role as a guild, providing thousands of apprentices the opportunity to discover and hone their skills in all aspects of broadcast production before moving on to more lucrative professional media positions.

Throughout the sixties and into the early seventies, WBAI opened its microphone to political agitators who were generally less concerned with the formal elements of broadcasting than with the righteousness of their message. Although these agitators often had important insights into the dynamics of American society, in their idealism (or dogmatism), they simply lacked the patience to master the skills of "good" radio. Unlike the first generation of "radio people," for whom the First Amendment was a cherished ideal, these newer broadcasters also tended to be less tolerant of diverse opinions.

(National Public Radio, which emerged at this time, also struggled to combine politically engaged reportage and commentary in a polished and professional manner. After struggling for a couple of years to provide a spectrum of opinion and experimenting with formal innovations, public radio acquiesced to a more polished, limited purview. Given its increasing dependence on government and, later, corporate funding, NPR rarely could offer the range of opinion, the fierceness of the polemics, or the experimental posture that could be found on Pacifica. The recent censorship of Mumia Abu Jamal's reporting is only the latest instance of NPR's basic failure to live up to its claims. Of course, many Pacifica alumni were among the first generation of NPR broadcasters. WBAI's Chris Koch was centrally involved in creating NPR's most important program, the evening news show *All Things Considered.*)

In Josephson's narrative, the fundamental problem facing WBAI in the mid-seventies was that the political activists ("radical lesbian ayatollahs," as he called them) had overwhelmed and eliminated the conscientious media professionals such as himself. While at times derided as a master of self-pity, Josephson was acutely conscious that during his tenure as station manager from 1973 to 1974, the subscription base was hemorrhaging. In the name of serving smaller, devoted audience niches, certain programmers refused to consider how their shows blended into a daily schedule. This was not a process that happened all at once. Nonetheless, by the end of 1974, Fass had gone on extended leave; Josephson was preparing to move to the Bay Area; and Julius Lester had departed, lamenting in his farewell note that WBAI was not able to model in its own station practices the sort of community it hoped its listeners would create.

Although it may be an exaggeration that "Larry had no politics at all,"[28] he was, by his own admission, not someone who wanted "to use WBAI to save the world." Some saw him more harshly, a typically arrogant male trying to hold on to his own power in the face of a rising feminist tide. He was clearly more impressed than many other staff members by the power of major fund-raisers on the board of directors, exemplified by their ability to help purchase the church building for the station on the strength of a promissory note. Josephson represented a minority who argued that development energy should go toward obtaining larger grants and major donor contributions, softening the station's reliance on its diminishing listener sponsorship.[29]

This funding strategy greatly impacted the facts of life at the station. On-the-air marathons might still glorify WBAI as "the voice of the movement," but the overall requirements of grantsmanship led to a greater stress on the "professional" and technical qualities of the programming to major donors. Little things around the station began to change. Tape-recording equipment, which had been available "for anyone who walked in off the street," was now locked up; volunteers needing mentorship came to be seen as hindrances to the older programmers. The overall ethos of the station subtly shifted toward enhancing the infrastructure.[30] Most consequentially, Josephson chose in 1974 to suspend publication of the *Folio* in a cost-cutting effort. This dramatic gesture indicated to some his elitist disdain for WBAI's core listeners.

The central issue for the board in the mid-seventies was WBAI's mediocre job building a multiracial audience. This particular problem led to their decision to employ new personnel and revamp the schedule. At WBAI, Julius Lester's *Uncle*

Tom's Cabin was a popular and long-running program highlighting public affairs and culture from a black perspective. As one of the country's best-known public figures in the Black Power movement, Lester's prominence at WBAI served to protect the station against charges of racism while he worked there. Other programming on Asian, Latino, and Native American issues was allotted regular slots at different moments.

Yet if Lester exemplified WBAI's commitment to racial diversity, to some it seemed that his programs were more of a piece with the idiosyncratic, personality-centered programming that the station had cultivated since the mid-sixties than with actual struggles taking place in Harlem or Bed-Stuy. In a telling example, one longtime listener recalls Lester less for his numerous programs on black life and politics than for his discussion of his conversion to vegetarianism and the moral dilemmas he faced at Thanksgiving dinner.[31] This anecdote indicates some of the problems the board faced in trying to come up with a strategy to expand WBAI's coverage beyond its diminishing white middle-class audience. (However one assesses WBAI's small audience base in nonwhite communities during this period, on the evidence of the archives and program guides, one can also argue that Pacifica has dedicated a higher percentage of airtime to civil rights, racial "minority" and indigenous people's movements, and Third World anti-imperialist struggles than any other broadcast network in the United States.)

CHAOS AND CRISIS

By 1975 the station was hardly thriving on chaos. Its debts were mounting; there was no consistency to the daily schedule; and there was a minuscule audience in the city's vast nonwhite communities. For the individual programmers, each hour remained a precious sanctuary. Their small audiences, after hours of on-the-air dialogue, had developed into intricate, passionate communities of listeners. In spite of the financial turmoil, many producers retained complete confidence that their *particular* program was essential in guiding New Yorkers toward the emerging ecological, gay, lesbian, Latino, and feminist movements that formed and reformed through the early seventies. Everyone acknowledged that there was a need for more Third World programming, yet daily airtime was limited and nobody was volunteering to cede slots for new shows. From the board's perspective, many shows on WBAI's schedule had by this point become little more than "vanity" programs.

Some staff, adopting the logic of commercial broadcasting, felt that WBAI's prime central position on the FM band was going to waste. One option floated at the time was to sell WBAI's prized channel for tens of millions of dollars and re-locate to another in the "left" end of the spectrum usually reserved for educa-tional and public broadcasting. This infusion of cash, it was argued, would have then enhanced the entire network, not just the station in New York. By the time this "solution" was suggested, no faction was in a strong enough position to ne-gotiate such a radical plan.

No doubt some of the difficulties WBAI faced in this period were generic to the challenges confronting alternative media (which at this moment meant pri-marily the press): as the sixties ended, how might nontraditional outlets represent the emerging political forces in a way that was both honest and partisan? More important, what could inspire the devotion of a creative staff willing to work long hours for little money in a period during which the animating inspirations of rev-olutionary anti-imperialism or the Age of Aquarius were on the wane? Given how quickly the new movements were factionalized along the liberal-radical axis, en-gaged reporters found themselves in awkward situations attempting to produce stories that would inform a larger audience without offending or betraying their activist friends. This conundrum, which first arose in the reporting on the Weather Underground and the Black Panthers, took on the quality of pitched battles at WBAI. As Paul McIsaac, a producer at WBAI, asked, should innovative, pioneer-ing gay programmer Charles Pitts be allowed to use his program to talk about "how wonderful it was to 'diddle' little boys?"[32] Should feminists so consistently promote a radical separatist line and deny men the right to participate in call-in shows, contradicting a central premise of "free speech radio"?

At the same time, the mainstream media were hiring more young, committed journalists, providing them with real salaries and larger audiences. It had become nearly impossible to attract skilled professionals to the contentious life of poverty that Pacifica promised.

Different observers lay the blame for the worst of the problems on a number of causes: the board's indecision over implementing new policies in the crucial period from 1975 to 1976, ceaseless rhetoric from the "sexual liberation front," Josephson's suspension of the *Folio* in 1974, the end of the Vietnam war. Undis-puted is the fact that between 1971 and 1976, the station had lost half its sub-scribers (from thirty to fifteen thousand), alienated many of its wealthy patrons, and was building up increasing debt, ultimately losing the church building in 1979 in a complicated tax case where the city revoked the station's educational status. No single person or ideology shoulders the entire burden for this. More than any-thing else, the waning of the sixties zeitgeist seemed to deflate the innovatory zeal WBAI had experienced for almost a decade.[33] Given these spiraling tensions, one might well wonder how WBAI managed to maintain itself without a full-blown crisis until late 1976.

In the fall of 1976, the board hired a new general manager, Anna Kosof; she in turn appointed a new program director, Pablo Yoruba Guzman. By most accounts, Kosof, Guzman, and their supporters on the board were unprepared for the intense turmoil this new management team unleashed. Even harsh critics concede that Guzman and Kosof did not create the problems they were hired to remediate. Nonetheless, it seems that this new management team, especially Kosof, was not "in sync" with the idiosyncratic culture of WBAI. Kosof had been a longtime organizer in community affirmative action and drug treatment programs; Guzman, a former programmer at WBAI, had been a highly visible and effective public relations coordinator of the Puerto Rican activist organization, the Young Lords, before becoming involved in the local salsa music business.

This new management made a conscientious attempt to transform WBAI explicitly into a "community station," defining community generally as the un- and underrepresented groups in the metropolitan area. Guzman's first (last, and only) order of business was his proposal for a completely revised schedule using daily salsa programming to draw in a larger audience, especially those potential listeners from New York's Third World community. Kosof, in her role as manager, struggled to remedy what she saw as rampant internal chaos with imperious directives ("No pot smoking in the employee lounge") and, far more consequentially, with injunctions to the staff to stop discussing internal station politics on the air.

Throughout the fall of 1976, the staff, both paid and volunteer, began holding ad hoc meetings to present a united front before the board in opposition to the impending schedule changes. Although they shared little genuine solidarity about larger programming goals or how to cope with the financial turmoil, the simple fact of the meetings, which originally drew between sixty and eighty participants, seemed encouraging. As an example of the staff's quandary, when Kosof offered to resign in November, many who disdained her rallied on her behalf, not because they believed she was well suited for her position, but because they wanted to prevent another torturous search for a general manager. The meetings, initially filled with hopeful enthusiasm, became increasingly agonizing; everyone realized that change was imperative, but no one seemed able to articulate the actual forms those changes should take.[34]

Guzman, more than Kosof, tended to be cast as the antagonist in the drama as it played itself out. Although WBAI surely needed someone experienced with affirmative action policies, Kosof was dismissed as someone simply too alien to the phenomenon of WBAI, "a straight, uptight woman who didn't have a drop of sixties blood in her body."[35] Guzman, however, was a far more logical choice for his position. A former WBAI programmer, movement activist, and dynamic personality, many originally believed he understood the nuances of station politics and might be able to meet the challenges he faced. His plan to institute a consistent daily schedule emphasizing music and other programming aimed primarily at the Latino and black communities was a plausible place to begin negotiations.

What seemed to truly rile the staff, however, was Guzman's arrogance and his insinuation that he had been hired to salvage the station from some old white hippies, accurate though his assessment might have been. He justified his proposal for more musical programming with the rhetorical, somewhat confusing claim that "this is not the sixties. . . . You can't just talk shit anymore. . . . I want professional revolutionaries."[36] Any attempt to modify his vision by other staff members was met by charges of racism. This race baiting by a respected nonwhite activist was highly provocative—if not to say "simply ridiculous,"[37] according to one opponent, who would have welcomed more Third World and salsa programs but not Guzman as program director.

In their defense, it should be said that Guzman, and to a lesser extent Kosof, wanted to bring to WBAI a reasonable model for Pacifica. They sought to revise the overall sound of the broadcast day based on the goals articulated for them by the board. Observers other than Guzman did see elements of racism in the staff's resistance to his plans. By the end of 1976, the hostility between staff and management and the rumors of the impending restructuring were a constant and bitter topic of programming. Finally, Kosof issued yet another memorandum calling for the cessation of the airing of the station's dirty laundry. Some of the staff refused; the memo itself became a topic of on-the-air discussion.

It was at this point that Guzman announced the full plans of his "nuevo barrio" schedule, promising to build the subscription base to fifty thousand members, people who were "dissatisfied by plastic radio, yet conditioned by it."[38] Rather than diffusing tensions, his plans only further riled the staff. In early 1977 Guzman held a public meeting where he presented his view of the situation. Some of the call-in and live radio shows would continue, but now in a much more tightly formatted overall schedule. In Guzman's words:

> The staff has had in the past several months de facto control of the station, but could not turn the situation around. Rather, decline and isolation accelerated. Thus, if left to their own devices, the present staff would take WBAI completely down

the drain.... this is the grounds from which spring charges
of racism, elitism, and block further progress.[39]

Ironically, although he had been hired to open the station up to Third World programming, Guzman fired the only black woman on staff, Deloris Costello, apparently believing her show to be implicated in the old regime. Rather than participate in further discussions, the staff viewed Guzman's plans as an effrontery and refused to negotiate. The station was "suffering from too much democracy," according to one board member.[40] The personalities involved made any real compromise impossible.

The strike, or lockout, began on 11 February, after Kosof told the board she no longer had control of the situation. At that point, the board decided to take the station off the air to diffuse some of the hostility. Before the power could be turned off, however, an ad hoc group of twelve announcers and engineers occupied the master control room at the station as well as the transmitter in the Empire State Building in order to broadcast their position one last time. For five hours, this band played music, presented their demands, and chatted with their audience before the transmitter was turned off. Some activists in the station stayed locked in for almost six weeks, surrounded and supported by listeners and other staff members.

The "union" insisted that the strike be seen basically as class struggle. Recalcitrant management was attempting to balance the books on the backs of employees while using the ideology of race to divide the staff. The union had two central demands: that paid and nonpaid staff be recognized as members, and that any change in programming be "consistent with Pacifica principles." New programs would be adopted only after discussions of an agreed-on timeline negotiated by the program director and the union. As one taped press release claimed:

> The format is going to change, we know that. But change
> has to be rational and planned... not en masse by people
> who have no experience in programming. People have to
> have opportunity to be creative without being crushed by a
> format.[41]

Hundreds of listeners formed "Friends of WBAI," picketed the homes of board members, demonstrated at the Empire State Building transmitter, and maintained a round-the-clock vigil at the station where some of the union members were locked in.

Discussions dragged on inconclusively. Initially the board argued that only paid employees could officially be union members, but since less than 20 percent of the workers were paid, this position proved untenable. (Precedence for accepting unpaid staff as union members had been set at KPFA in Berkeley in early labor disputes.) Although the union "agreed" to let the board retain nominal control over programming, the board agreed not to implement Guzman's proposals. Claiming victory, the union returned to work, appropriately enough, on April Fools' Day, 1977.

By the time the strike was over, Guzman was gone, followed by Kosof a year later. George Fox, the most influential member of the local board who had the trust of some of the staff members, recalls that "the crowd was a lynch mob, using microphone wire to hang Kosof."[42] Most of the renegade programmers signed a letter promising to obey all FCC policies in the future — that is, not usurp the transmitter and broadcast studios. After a reprimand, most were allowed to maintain their shows. As Celeste Wesson, one of key organizers of the union, put it poignantly, "We saved the station from the board, but we couldn't save it from ourselves." No one felt any elation at the "victory," and the station "seemed like a morgue."[43] Some activists scarred during this period, most notably Bob Fass, have come to feel they have "sacrificed [their] best years to WBAI"[44] and now wonder about the cost. Guzman's departure led to more than a decade in which "minorities" would have a very limited voice in the daily sound of WBAI. By Reagan's election, almost all of the staff involved in the "free radio" had left the station, driven by economic necessity and choice into other careers.

> Now, in the 1990's several million people may listen to one of
> my [NPR] reports and I might not receive a single response.
> But in the 1970's, on that 5:00 A.M. show, it was possible to
> create a community of listeners. . . . At five in the morning
> there is nothing that can't be changed.
>
> — MARGOT ADLER, *HERETIC'S HEART: A JOURNEY*
> *THROUGH SPIRIT AND REVOLUTION*

A student of broadcasting history might wonder what James Rorty, vociferous
critic of early corporate media, would have thought of the crisis at WBAI. In the
early 1930s, educators, civic activists, and church leaders watched in dismay as
the Federal Radio Commission stripped the broadcast licenses from their stations
and, under the rubric of "public interest," gave them to commercial broadcasters
(see chapter 1). In meeting after meeting organized by the National Council on
Educational Radio, the noncommercial operators gathered to address this baleful
situation. They consistently charged that commercialization would debase Western literary and musical heritage: "Private radio monopolies deriving their revenue from advertising [are] dead set against the fundamental ideas which underlie modern civilization."[1] Noncommercial broadcasters argued for the Arnoldian
high ground, hoping to retain their licenses under the mantle of providing the
public with programs of cultural excellence, much as PBS does to this day.

Into this fray jumped James Rorty, the former ad writer turned socialist. Rorty
had for years been an acerbic observer of the commercial media and was in general alliance with the educators. However, at the suggestion of serving the public
by providing works of canonical excellence, he rebelled:

> Concerning the concept of culture, the point should be made
> that we do not have in this country *a* culture. When we say
> "we" what we mean is the particular group with whose in-
> terests we find ourselves identified. . . . We have a fragmented
> civilization, with not one culture but many cultures and many
> definitely conflicting interests . . .[2]

In the midst of a debate about the true meaning of broadcasting in the "public interest," Rorty insisted that it was the very premise of a single interest that was misguided.

It was not until the transformation of Pacifica in the late sixties, when stations in New York, Los Angeles, and San Francisco struggled to provide various groups with regular programming, that any broadcast institution in this country began the as yet unrealized task of accommodating Rorty's radical cultural pluralism — of confronting the massive social and cultural fact that "when we say 'we' what we mean is the particular group with whose interests we find ourselves identified." It would be stretching the truth to claim that Pacifica negotiated the situation with no problems. Opening the ether to "particular groups" ceaselessly testifying to the fact that "we have a fragmented civilization" was often torturous, accomplished only at the cost of work stoppages, deep animosity, and lingering distrust.[3] At the same time, WBAI's evolving relationship with two forms of community, one vast and heterogeneous, one smaller and more uniform, provides different models of how the media might serve and participate in the utopian project of democracy.

The previous chapter charted distinct moments that WBAI laid the groundwork for community radio. The first was the interaction of *Radio Unnameable* and other live radio programs in the mid-sixties with the vast, inchoate counterculture in the New York metropolitan area. Binding the programs and audiences together was the project of ending the Vietnam War, but combined with this manifest political task was the Whitmannian conviction in the power of the imagination to transform existence altogether. The second moment, evolving from the first, began when the daily schedule increasingly became a locus for smaller groups, bound together by webs of curiosity, loyalty, and love, to use the microphone to call out to each other and initiate a conversation on quotidian matters of abiding personal interest.

Broadcasts addressing the division of labor in the household, or the demand for two men to feel free to hold hands in public, clearly grew out of, but also differed from, the *Radio Unnameable*–yippie synthesis of the antiwar and counterculture energy. Stepping back to consider more theoretically the transformations of programming helps place WBAI's experiment within the overall changes occurring in the politics and culture in this period.

Pacifica's radicalism, as it crystallized at WBAI, was an emerging synthesis of elements, not all of which were precisely anticipated by Hill. *Radio Unnameable* and the free radio that followed used the evocative power of radio to promote an iconoclasm toward all that is, broadcasting a continuous proposal that no boundary was sacred; any media format, cultural norm, or political conviction could, indeed must, be open to collective revision. The programs provided a paradigm of what they demanded in other spheres. Fass mastered radio's power to spur playful imagination and modeled the ways in which this imaginative energy might be deployed toward concrete political activity. In the early morning hours after most people had gone to bed, the counterculture in New York used listener-supported radio to enact the whimsical, polymorphous, and revolutionary character of the moment, one in which fantasy was unleashed, permeating public consciousness and political praxis as a whole.[4]

Community, based on creating collective events, was typified by ad hoc "happenings" such as the fly-in, where "even though we had never met, because we were all connected through Fass we felt connected to each other. It was like being part of the greatest party you could imagine." This is not a community based on propinquity and ongoing interpersonal interaction; yet for participants, the sort of affinity such events produced stood for community — for being involved with others engaged in a common project based on shared ideals and enthusiasms. At what level do the media operate in such a situation? On the one hand, the very success of such an outlandish project might raise the specter of the discredited "hypodermic needle" model of broadcast propaganda, in which the media is seen as capable of transforming opinion and behavior with the wave of its magic ethereal wand. On the other, this symbiosis between program and audience might also be a glimpse of "undistorted communication"; it anticipates, before its arrival, a form of media carnival that a more egalitarian and playful society might enjoy.

By the mid-sixties, FM radio's overall soundscape had internalized an enormous amount of the joie de vivre of *Radio Unnameable* and a significant portion of its politics. It would be folly to claim that the entire cultural politics of FM radio in the sixties derived from Bob Fass. Nonetheless, Fass, and the free radio at WBAI he inspired, played a vital leadership role in this period, inspiring such legendary DJs as Tom Donahue to decorate stations with flags from the Vietcong. Although Donahue, "the father of FM radio," deserves credit for expanding the variety of rock music programming available over the airwaves, media scholar David Armstrong notes that in his overall project, Donahue was essentially following "the free form pioneers on Pacifica."[5]

Once the larger mythical "movement" began to splinter, and with it the waning of belief that prayer and incense could levitate the Pentagon, much of the radical and alternative media were left without a base — the "underground" press either surfaced or died. Commercial forces and formats seeped into FM radio. PBS, severely attacked by Nixon, was neutered as a critical force, turning to greater corporate sponsorship for its imported "masterpieces."[6] In this situation, WBAI, and Pacifica overall, desperately sought other options. The producers and audiences for the feminist, Native American, and Earth Day shows of the early seventies, while still intimately connected to the legacy of the antiwar and civil rights movements, demonstrated that both "the" movement and "the" system were not monolithic phenomena. Meiklejohn's version of freedom of speech's relationship to democracy was turned inside out: in these programs, the core issue was not that everything *worth* saying be presented but that everyone wishing a forum be given an opportunity to speak. Who might be able to estimate the "worth" of a given program became an unresolvable problem, one whose thorny legacy continues to throttle the left and libertarian impulse to universal ideals to the present.[7] (Whereas some Marxists at all stations would claim that proletariat consciousness should be the ultimate arbiter of political legitimacy in these circumstances, their voice was generally muted by the peculiar behavior of the Weather Underground and other "revolutionary" tendencies of the period.)

For station management, coping with the range of new groups in the early seventies agitating for programming was surely vexing. To many in both the station and the audience, the broadcast day was increasingly filled with confusion, sloppy programming, and vituperative personal attacks. Yet rather than give in to forces of bureaucratic professionalism, as did NPR, or commercialism altogether, as did much of FM radio and the underground press, WBAI met the problems head-on, offering a unique, if problematic, model of station democracy. From 1969 to 1976, as the ideals and energies of the counterculture and antiwar movement seeped away, WBAI persisted in the struggle for a new society, enabling tens if not hundreds of thousands of people to build a meaningful vocabulary for solidarity. From the perspective of 1967 or 1997, the accomplishments of this period might seem a banal or compromised achievement. Yet at the moment in which it emerged, "community radio" provided a vital, novel opportunity for defining and contesting identity, affiliation, and strategy.

Rather than the protest and radical agitation that marked the antiwar movement and its programming on WBAI several years earlier, in the early seventies, announcers chose a different path. A radio station that for years had celebrated frank, uncensored call-in shows, personal expression, and political agitation now also opened its microphone to a more intimate and personal dimension. Believing that sharing life histories was an essential step in raising consciousness, new programmers were inventing a politics for which they had few precedents. Public, voluntary self-exposure for the purpose of stripping stereotypes of their force was a heroic act of a different order, lacking the theater of the civil rights and antiwar movements, but demanding an inner resolve and courage that utterly transformed those who took the risk.

As Nancy Fraser argues, the very fact of opening "private" realms to "public" dialogue constitutes a necessary first step for counterhegemonic praxis. Although "breaking out of discursive privacy" is "slow and laborious," "when this happens, previously taken for granted interpretations ... are called into question, and heretofore reified chains of ... relations become subject to dispute."[8] Using their broadcasts to "break the discursive privacy" in which hierarchy thrived, programmers enabled questions of intimacy, sexuality, family, and cultural identity to have an extensive, consistent, and uncensored airing. In this way, WBAI's programming served as another approximation of how a democratic media might lay the groundwork of new communities of interest and struggle.

What, after all, is "community"? In the words of Josiah Royce, the philosopher who coined the term "beloved community" and spent much of his life charting the conditions for modern affiliation:

> Men do not form a community ... merely insofar as [they] cooperate. They form a community when they accompany this cooperation with that ideal extension of the lives of each member, ... [who] says: "This activity which we perform together, this work of ours, its past, its future, its sequence, its order, its sense, all these enter into my life, and are the life of my own self writ large."[9]

In a theory drawing heavily on Pauline theology and Deweyan pragmatism, Royce argued that it was not geographical location but collective "ideal extensions" toward a common past (crucifixion) and anticipated future (resurrection) that creates community. Royce's communitarian ethos synthesizes three elements: memory, hope, and collective practices. These constellate as the conditions for loyalty among comrades, "loyalty" being for Royce the means and end of community life. Royce argued that the solution to the problem of community was in shared interpretation of a common history. This narrative enabled those sharing it to participate in collective projects building toward an idealized future.

It is with Royce in mind that we might now consider how the evolution of programming at WBAI serves as a microcosm for political transformations that closed the "moment" of the sixties—that complex "common objective situation" beginning in the mid-fifties with the civil rights movement and the Cuban Revolution and ending conclusively, for WBAI, with the strike in 1977.[10]

The shows that emerged later in the decade, while still connected to the vitality of the antiwar and civil rights movements, served to parse the monolithic image of "the system" or "U.S. imperialism" into more concrete targets. In the early seventies, a host of new forces groped toward an understanding of the radically contingent and partial nature of transformational politics. They struggled to construct a usable past, weaving together thousands of personal testimonials, using the feminist consciousness-raising experience as the paradigm. In this context, a regularly scheduled program on WBAI became a central means for arguing, exploring dreams, and, most important, reaching out to others to produce a coherent, if partial, narrative with which to generate "this activity to perform together."

That struggles over social power are "radically contingent" has become a central tenet of oppositional politics within the postmodern terrain.[11] The programming at WBAI in the sixties and seventies exemplifies in concrete form how this elusive category works. When Fass and the yippies questioned the fundaments of bourgeois life, they located the center of that life in the public world: airports, stock exchanges, presidential conventions, the Pentagon. Fass's energies went into transforming these common places of "Establishment" power into vast spectacles, holding them up to ridicule. How had any of these things earned our respect? he asked, and by asking, every night for years, *Radio Unnameable* used radio to erode the prestige and respect these institutions held.

(To be sure, exposing the historically unjust conditions on which prestigious institutions maintained their authority was only the first step in the process. For many idealistic youth pitched into the battle against the dominant powers at this moment, there was very little understanding of the enormous resources, coercive and attractive, that "the Establishment" had on hand. Although pajama-clad peasants seemed to be battering history's most powerful military force, this image was subject to serious misunderstanding. At no point in this period was there a revolutionary situation in the United States, something theorists such as Herbert Marcuse spent years warning against, often in programs broadcast on Pacifica.)

While not ignoring these major sites of political struggle, the next generation of programmers insisted that basic changes within far more intimate realms were both necessary and revolutionary. Participants recall these shows as ones in which thousands of people explored the nature of personal relationships with the same fervor that earlier shows had attacked the war. Why, for instance, should the paradigm of human intimacy be chained to the heterosexual norm of marriage and the nuclear family, itself a historical contingency whose prestige has been fiercely guarded by patriarchal authority for millennia? This search for new definitions and objects of affection rippled into the larger questions of general values worth committing oneself to, so reminiscent of Hill's earliest roundtables for KPFA. When intimate life became the site of public dialogue and struggle, novel and at times dramatic forms of self-exposure, transformation, and solidarity occurred. ("At five in the morning there is nothing that can't be changed.") Brazilian author and activist Paulo Freire highlights the liberating subjectivity that emerges in the moment when the "oppressed" learn to speak for themselves and take control of their lives, a process that takes place over a long period of education and dialogue, not in a punctual, violent upheaval. The programs on WBAI in the early 1970s helped listeners begin the "dialogic" process wherein they found their own voice and discovered their capacity for agency.

As chapter 6 described, WBAI not only reflected the hopes and accomplishments of this moment but also participated in the intense factionalization of the newer social movements themselves. In a sense, one cannot separate the failures from the successes. The very passion and vigor that opened new spheres of the self to media dialogue also diminished the patience necessary for the lessons to be digested. With only the first steps taken toward a shared interpretation of a common past, with neither the issues nor the strategies fixed by any external, universal set of values, with little consensus on who were allies or enemies, the ad hoc coalitions formed in the midst of struggle rarely achieved an abiding solidarity. Rather, much energy poured into exclusionary forms of identity based on facile notions of "authenticity," in which "the only common condition worth thinking about was the impossibility of commonness."[12] It is for this reason that so many contemporary critics cast a baleful eye on this moment of the birth of "identity politics." As consistent fissuring inside these groups occurred, the programs themselves highlighted the great difficulty that apparently coherent groups such as "feminists" had in creating a common memory or community — "white" feminists, dykes, older women, married women of color, all "remembered" and responded to different aspects of patriarchy. Not unreasonably, one might have hoped that the common use of the scarce airwaves would have helped to build consensus and begin to synthesize these disparate "identities." This, after all, was Hill's original hope: using radio for finding common ground among the variety of pacifists with whom he worked.

Although some unity surely emerged in the genesis of community radio at WBAI, neither the staff nor the programs themselves adequately resolved the conflicts generated by the diversity they celebrated. Rather than only expanding the boundaries of the public sphere through the assertion of the rights of this or that particular group, or struggling in the name of individual conscience against unjust state policies, the most interesting and contentious of these programs, like the postmodern political movements from which they derived, forced people to "react, respond, sometimes to think. . . . Change in this sense is a bumpy process."[13]

Yet if the birth of community radio came at a cost, it is also clear that Pacifica did transform itself, welcoming many groups to its microphone and allowing them autonomy in programming. In choosing this path, the network modeled an almost unprecedented use of the media, new even by Pacifica's own standards. From this relatively unmediated public access to the airwaves, and from the discussions, controversies, and consciousness-raising that this programming generated, emerged a form of broadcasting now known as "community radio." Seen by its present-day partisans as a "genuine communication tool that encourages creativity and allows popular access," community radio encourages "expression and participation" on the part of marginalized groups previously "without a voice."[14] Around the globe,

> women, indigenous peoples, ethnic and linguistic minorities, youth, the political left, peasants, national liberation movements, and others are discovering radio as a means of political and cultural intervention and development. They are transforming radio into a medium that serves their needs — a medium that allows them to speak as well as to hear.[15]

Although Pacifica has not been the only model that community radio has followed, the network has remained one of the oldest and most prestigious examples for several generations of independent broadcasters internationally.

In conclusion, let us return to Vera Hopkins and her belief that "community radio" was counter to Pacifica's original goals. Intermittent sloppiness and unabashed dogmatism may very well have offended the aesthetic and political sensibilities of Hill and his cohorts had they been able to tune in in 1972. Nonetheless, it seems more apt to acknowledge community radio as one potential outcome of the dynamic ideals infusing the network from the start, fundamentally connected with the original injunction to produce programs to diffuse social antagonisms.

Consider in this regard Gordon Wood's discussion of the democratization of mind in the American Revolution. Wood lavishes praise on the Federalist accomplishment: the creation of a new nation based on untested principles of representative government. Yet in their very brilliance, men such as Madison, Adams, and Hamilton were not themselves representative of the people as a whole. The Federalists, Wood argues, created the circumstances for a new form of sovereignty — democracy. Paradoxically, in doing so, they had elaborated conditions for a political regime from which they had by and large eliminated themselves. With the writing of the Constitution and the disputes over its ratification, the founders established a new political culture altogether, one where

> truth was actually the creation of many voices and many minds, no one of which was more important than another and each of which made its own separate and equally significant contribution. Solitary individual opinions may thus have counted for less, but in their numerous collectivity they now added up to something far more significant than had existed before.[16]

This well describes the process that occurred in the creation of community radio at Pacifica. The first generation of programmers held a deep faith in the potential of each person to act conscientiously and creatively if guided by men and women who themselves modeled such behavior. Radio would be the tool to promote and distribute this intelligence. Then, in the wake of the civil rights struggle and the movement to end the Vietnam War, a range of new groups demanded public recognition in terms that they chose for themselves. Following the network's original injunction to serve the causes of peace and study the sources of social strife, Pacifica made itself available to these new actors on the social scene, in the process undergoing a radical, at times painful transformation, a "democratization of the mind."

> You can, nevertheless, work toward a situation that keeps alive
> the power to break the limits: to think thoughts that shatter the
> available canon of reason and discourse, to experiment with
> forms of collective life that the established practical and
> imaginative order of society locks out or puts down.
>
> —ROBERTO UNGER, *PASSION: AN ESSAY ON PERSONALITY*

This book has not attempted to disguise its admiration for Pacifica's accomplishments. Neither has it narrated a triumphalist version of Pacifica's history. Little in Pacifica's opening the airwaves to controversy, erudition, and diversity has been simple. Lack of financial support, internal political and personal struggle, and constant surveillance by political enemies, both within the government and without, have marked Pacifica's history. Yet in the past five decades, oppositional social movements, cultural avant-gardes, and various alternative media have come and gone while the radio network remains and continues to evolve. The fact of its persistence against many odds may be attributed to Hill's initial understanding of the special reciprocal responsibility established between the volunteer and poorly paid staff and the network's hundreds of thousands of listener-subscribers, a reciprocity impossible within the world of commercial broadcasting.

As has been the fate of radical libertarian politics generally, the network has consistently seen its utopian aspirations falter on the shoals of both human foibles and repression from the forces of order. A cynic, reviewing Pacifica's history, could note that there is little consensus of what democratic broadcasting might communicate, which audiences it should serve, and the manner by which it could sustain itself. That free speech on the airwaves is the cry of Howard Stern and Rush Limbaugh, that human creativity is ever more rather than less tightly bound to the imperatives of the market, that economic disparities are greater today than fifty years ago, and that meaningless slaughter persists throughout the globe might elicit an even greater level of pessimism.

How does one finally gauge the accomplishments of a project whose lofty goals remain unmet or have been distorted beyond recognition? There is an important, well-rehearsed answer that the Left provides. Beginning with Marx's "On the Jewish Question," historical materialists have criticized those who elevate "civil" liberties. This critique is based on the fundamental recognition that within a capitalist democracy, actual freedom is always and ultimately constrained by that locus of unfreedom, the market itself. Corporate capital might allow, indeed even encourage, consumers to pursue "expressive" freedoms in every arena except in the sphere of daily work activity—where one spends most conscious, productive time. From the early Marx to the late Marcuse (a frequent and charming guest on dozens of Pacifica broadcasts), socialists have warned against liberal, or reformist, politics that privilege civil rights, personal transformation, nonviolence, or community over the struggle to unite forces engaged in class conflict. Postmodern social movements espousing identity politics are simply the latest manifestation (and failed instance) of an idealistic project gone astray.

It is also possible to consider the struggles Pacifica endured as indicative of the antagonisms internal to the ideals of liberal democracy itself. The persistent tension between equality and autonomy within the United States, as Tocqueville observed, remains a defining circumstance of democracy. Within the network's development, one pole (justice, equality, community) has consistently interacted and struggled with the other (excellence, liberty, autonomy). Therein, perhaps, lies a different way of understanding Pacifica's ongoing, uneasy, at times bitter transition from "free speech," to "free," to "community" radio.

Consider in this regard Ed Goodman's (WBAI's general manager in the early seventies) summary of the lessons of this period. In a program guide from 1973, he lamented about the persistence of strife at the station:

> The tension between access [to the microphone] and quality appears to me inevitable. That tension is now more pronounced due to the heightened consciousness of various disenfranchised groups such as gay people, blacks, women, etc. The fact is that when access is first enjoyed by any previously denied group, pain, anguish, and anger are the main ingredients that come across the air.... This is not a tidy process. It is cumbersome and replete with loose ends, dangling participles, false starts, and effrontery.[1]

This may be one of the central lessons of the network's history and one of the central riddles: Must "access," that fundamental egalitarian ideal, interfere with "quality"? In the yearning for both justice and excellence, who is to determine that Archimedean point of balance? There is also a simpler point that Goodman's lament — and Pacifica's history generally — starkly brings to mind: opening the public sphere to more voices is not a "tidy process."

A fair question to ask is whether it should be. Indeed, one might argue that Hill anticipated these very struggles when he called Pacifica a "brash experiment." An experiment is not a neat, predetermined sequence of events, but a risky venture subject to error and failure. Justice Holmes ("life itself is an experiment") would surely acknowledge that much experience is "replete with loose ends, dangling participles, false starts, and effrontery." Nonetheless, the experimental disposition also keeps imaginative alternatives in mind, enabling both innovation and solidarity. It provides for the plasticity of intellect within democracy.

Had Pacifica's project simply been democratizing the public sphere with "broad, wide, varied, and rich" programming that questioned the political status quo, or providing outcast groups with unmediated access to the airwaves, its accomplishments would surely merit attention. That the process was filled with trial and error and fraught with difficulty should only be expected of a grand experiment. However, as the original program director, Eleanor McKinney, insisted, the heart of Pacifica was something more than this.

> Pacifica was not "political"; it was human. . . . Peace was the real meaning of Pacifica. It meant let us reason together and explore together and even have compassion for each other in the process.[2]

If "peace was the real meaning of Pacifica," its "brashness" appears, on the surface, to undermine the "compassion for each other" that McKinney invoked. Yet, in the end, this swagger was less an impediment than a necessary complement to Pacifica's pacifism: using the airwaves to eliminate war and the occasions of war would be a dynamic struggle, not a languid "world of clerks and teachers, of co-education and zoophily, of 'consumers leagues' and 'associated charities.' "[3]

Pacifica grasped the central element in James's critique of pacifism: a moral equivalent of war could not come simply from seminars, or conferences on international law and trade, or prayer and homiletics. War grips the human imagination with the horrible thrill of participating in a massive, brutal contest of outinjuring, played for keeps in that "supreme theatre of strenuousness," the battlefield.[4] Pacifism must demonstrate equivalent brashness to galvanize the passions. An alternative to war needs to elicit the same blend of ingenuity, courage, and selflessness as combat itself. It must engage an audience willing to heap honor on the participants. The mass movements led by Gandhi and King well exemplify the heroic nonviolent activity that James invoked—projects that involved not simply the intellect but the souls and bodies of the recruits in vast public spectacles.

Pacifica's ideological origins lay in this dynamic strand of the radical pacifist movement, and its practices remain tied to the revolutionary lineage of the First Amendment. By broadcasting "to expose lies and falsehoods," "to avert evil by processes of education," and "to free men from the bondage of irrational fears,"[5] Pacifica followed Brandeis's vision in its path toward a moral equivalent of war. It accepted the basic, intuitive truth that violence and aggression stem in part from the failure of dialogue to resolve disputes. Given enough time and opportunity to speak frankly about differences, both persons and nations would not resort to slaughter to make their points.

However, war should not be viewed simply as a failure of politics or dialogue; as James insists, the very violence of battle also exerts a positive attraction on the imagination. A second component of Pacifica's libertarian vision addressed this issue. For Brandeis, "the freedom to think as you will and speak as you think" is the mark of "courageous, self-reliant men," who have "confidence in the power of free and fearless reasoning."[6] Note the pattern of adjectives: "courageous," "self-reliant," "free," and "fearless." For Pacifica, the engaged public participation of citizens in affairs of state (or community) provides the vital, self-transcending activity necessary to sublimate and replace the attraction of violence. Can passionate partisan debate over culture and society sustain the "martial virtues" of honor, tenacity, and ingenuity without "the depth, massiveness, intensity, and speed of injuring that is central" to war's activity?[7] At the moment, these questions cannot be answered. Neither, however, can they be avoided.

Since KPFA's first transmissions in 1949, Pacifica's staff has operated with a singular insight about the relationship of the mass media to the social world at hand, one that resonates dimly, if at all, in the current world of broadcasting—commercial or "public." They believed and acted as if they were part of a larger utopian process in which radio would be used to eliminate social antagonisms and promote creative expression. "Isolated from those to be communicated with, shut off in a tiny room absent of living matter, staring at clocks, meters, and machinery, pushing buttons, pulling levers, surrounded by glass and metal," Pacifica's programmers used their patched-together equipment to "communicate experience and feeling."[8] "Free form radio," wrote Julius Lester in 1974, "requires that the producer be open and vulnerable to human experience and that his show speak from his own condition as a vulnerable human to that same condition in his listeners, because that is the basis of existence which all of us share."[9]

Lester's recasting of Hill's sophisticated, morally responsible announcer who claims, "Hey, man, this is the way it is. Listen to it," into an "open and vulnerable" human seems a plausible frame for the history of Pacifica narrated in this book. It provides a means of appreciating the network's evolving synthesis of pacifism, expression, and community. Pacifica has provided much airtime to persons uniquely able to acknowledge, indeed celebrate, their own human frailty — "the basis of existence which all of us share." Lester, and Pacifica institutionally in its first three decades, had not fallen victim to contemporary cynicism. When Hill spoke of "morally responsible" or Lester of "open and vulnerable" programmers, they were speaking authentically, as idealists knowing in both their hearts and their minds that the media, unchained from the profit motive, were capable of promoting wonderful transformations, that broadly communicating the values of moral responsibility and vulnerability was possible, and necessary, for a world without war to come into being. They modeled in their programming the ideals they professed.

This experiment in the broadcasting of "vulnerability" is another, considerable aspect of Pacifica's project. On the most basic level, acknowledging the fact of personal vulnerability (and its corollary, mortality) serves to loosen the irrational vehemence behind the desire to annihilate an enemy or harm another. At the same time, it responds to the call from the early critics of commercial media — how can radio serve to maintain the plasticity of intellect necessary to combat the pull of habit?

That "no aspect of character is safe from being transformed"[10] is, as Lester suggests, an opportunity for learning and growth. As cast by legal theorist Roberto Unger, accepting human frailty and incompleteness "frees you from a shallow and constraining view of who you are. . . . You learn to experience yourself as an identity that is never wholly contained by a character . . . and that grows by the willed acts of vulnerability that put a character under jeopardy. . . . You accept jeopardy as a condition of insight."[11]

Pacifica has struggled to embrace this dialectic of jeopardy and insight inside its programming practices. The corporate takeover of the media more than six decades ago foreclosed on the possibility that these challenges would have a place in the universe of broadcasting (outside of game shows). Largely undistorted by the interference of the market or the state, Pacifica retains to the present a rare freedom for a broadcast outlet. Its risk taking in the pursuit of insight and innovation rather than profit serves as a model for struggles for social justice and democratic media. This book has highlighted some of the central points of jeopardy — the firing of Hill, the Carlin case, the strike at WBAI. It has also discussed some of the most lasting accomplishments—the initial defiance of McCarthyism, the "free radio" of the sixties, the ceaseless commitment to political and aesthetic radicalism of various stripes.

Rudolf Arnheim suggested in the 1930s that radio could "draw the listener's attention to the expression and content of much that he ordinarily passes by with deaf ears." It was to capture this attention and turn it toward the service of peace that Pacifica was founded. Using radio's unique capacity to harness the power of the spoken word—the medium of consciousness itself—the network has struggled to repair the damage that war, a culture of consumption, and commercial media have wrought upon our national psyche. Alone Pacifica could not transform the ever-deepening impersonality of contemporary life, the banality of consumer culture, the persistence of war, and the erosion of community. Rather, as a witness, commentator, and actor, its overall programming provides an ongoing chorus of voices, calling to mind an ideal of a peaceful, democratic, global community yet to be realized.

PREFACE

1. Erik Barnouw, *A Tower in Babel,* 96.

2. U.S. Congress, Senate, W. G. Cowles letter to Hon. Hiram Bingam, discussion of the Proposed Federal Radio Act, 67th Congr., 1st sess., *Congressional Record* (1 July 1926): 12500.

3. Martin Aylesworth, "National Broadcasting," 27.

4. *EXTRA!: The Magazine of FAIR* 7, no. 5 (January–February 1994): 5.

5. John Dewey, *Freedom and Culture,* 185.

6. "Prospectus for Pacifica Foundation, 1946," in *The Pacifica Radio Sampler,* ed. Vera Hopkins. This nearly seven-hundred-page collection of chronologies, articles, internal memorandums, letters, and some of Hopkins's original notes is unpaginated. When I refer to a specific article that has page numbers, they will be given. Hereafter many of the references to this compendium will simply be cited as *Pacifica Radio Sampler.*

INTRODUCTION

1. S. E. Frost Jr., ed., *Education's Own Stations.*

2. Robert W. McChesney, *Telecommunications, Mass Media, and Democracy.*

3. Walt Whitman, "The Real War Will Never Get in the Books," 779.

4. "Articles of Incorporation," Pacifica Foundation, 19 August 1946, in *Pacifica Radio Sampler.*

5. Ibid.

6. Steven Schiffrin, *The First Amendment, Democracy, and Romance*, 87.

7. Alexander Meiklejohn, *Political Freedom*, 86–87.

8. Saying this in no way indicates that current participants in Pacifica's experiment are any less skilled, committed, or courageous than earlier broadcasters.

9. Oliver W. Holmes, *Abrams v. U.S.* (250 U.S. 616, 624, 1919), quoted in Lerner, *The Mind and Faith of Justice Holmes*, 312.

10. Dewey, *Education and Democracy*, 86–87.

11. Frost, *Is American Radio Democratic*, v.

12. George Kateb, *The Inner Ocean*, 44.

13. Walt Whitman, *The Works of Walt Whitman*, 221.

1. THE RISE OF CORPORATE BROADCASTING

1. Erik Barnouw, *A Tower in Babel*, 55.

2. Susan Douglas, *Inventing American Broadcasting, 1899–1922*.

3. Quoted in Werner Severin, "Commercial vs. Non-commercial Radio during Broadcasting's Early Years," 494 (emphasis added).

4. James G. Harbord, "Radio and Democracy," 215.

5. David Sarnoff, *Looking Ahead*, 48.

6. William Peck Banning, *Commercial Broadcasting Pioneer*.

7. Barnouw, *Tower in Babel*, 108.

8. Banning, *Commercial Broadcasting Pioneer*.

9. Merlin Aylesworth, "National Broadcasting," 27.

10. Frost, *Education's Own Stations*.

11. House Hearings discussing the Proposed Federal Radio Act (emphasis added).

12. John Dewey, *The Public and Its Problems*, 126.

13. Listed in *A Thirty-Year History of Programs Carried on National Radio Networks in the United States, 1926–1956*, ed. Harrison B. Summers.

14. McChesney, *Telecommunications, Mass Media, and Democracy*, chapter 2.

15. E. Pendleton Herring, "Politics and Radio Regulation," 173.

16. Ibid., 174.

17. Jimmy Morris, *The Remembered Years*.

18. "National Committee on Education by Radio," 3.

19. Llewellyn White, *The American Radio*, 108.

20. National Association of Broadcasters, *Broadcasting in the United States*.

21. Quoted in McChesney, *Telecommunications, Mass Media, and Democracy*, 208.

22. NACRE, *Four Years of Network Broadcasting*.

23. Ibid., 58, 72.

24. Summers, *A Thirty-Year History*.

25. James Rorty, *Our Master's Voice*, 270.

2. LEW HILL'S PASSION AND THE ORIGINS OF PACIFICA

1. Emile Arnaud, quoted in Roger Chickering, *Imperial Germany and a World without War*, 60.

2. Karl Marx, *Capital*, 926.

3. Karl Schmitt, *The Concept of the Political*.

4. Chickering, *Imperial Germany*, 34.

5. William James, "The Moral Equivalent of War," 1285.

6. Ibid., 1288 (emphasis added).

7. Ibid., 1287.

8. Keegan, *History of Warfare*, 355.

9. Zechariah Chaffee, *Free Speech in the United States*, 3.

10. Samuel Walker, *In Defense of Civil Liberties*.

11. Peter Brock, *Twentieth-Century Pacifism*, 41.

12. Charles Chatfield, *For Peace and Justice*, 106–7.

13. Roy Finch, in "Nonviolence in a Violent World," KPFA, Pacifica Foundation, 1965.

14. Quoted by David Dellinger in the foreword to *Against the Tide*.

15. Lawrence Wittner, *Rebels against War*, 3.

16. Cynthia Eller, *Conscientious Objectors and the Second World War* (emphasis added).

17. Niebuhr, quoted in Eller, *Conscientious Objectors*.

18. Eller, *Conscientious Objectors*, 50–51.

19. Roy Kepler in Dellinger, *Against the Tide*, 24 December 1984.

20. Jim Tracy, "Forging Dissent in an Age of Consensus."

21. Ibid., 95.

22. Ibid., 95–96.

23. Ibid., 98.

24. Ibid., 176.

25. Ibid.

26. "NSC-68: A Report to the National Security Council," 385.

27. Tracy, "Forging Dissent," 177.

28. Ibid., 180.

29. Appendix to "Report to the Executive and Advisory Members of Pacifica Foundation."

30. Kenneth Rexroth, *An Autobiographical Novel*, 519.

3. LISTENER-SPONSORED RADICALISM ON KPFA

1. Edwin Nockels in Robert McChesney, *Telecommunications, Mass Media, and Democracy*, 76.

2. Carolyn Marvin, *When Old Technologies Were New,* 211.

3. Erik Barnouw, *The Golden Web,* chap. 3. See also J. Fred MacDonald, "Government Propaganda in Commercial Radio."

4. "Birth of Pacifica," WBAI, Pacifica Foundation, 1966.

5. "Articles of Incorporation," Pacifica Foundation.

6. "The Exacting Ear," WBAI, Pacifica Foundation, 1966.

7. "Report to the Executive and Advisory Members of Pacifica Foundation."

8. The term used by Hill for KPFA during an on-the-air fund-raising pitch in 1956.

9. "Kenneth Rexroth Remembrance," KPFA, Pacifica Foundation, 1982.

10. "Playing in the FM Band," WBAI, Pacifica Foundation, 1974.

11. "Rexroth Remembrance."

12. Hill, "The Theory of Listener-Sponsored Radio," 24.

13. Ibid.

14. McKinney, introduction to *The Exacting Ear,* 10–11.

15. Hill, "What Is an Audience," 25.

16. Lewis Hill, *Voluntary Listener-Sponsorship,* 57–58.

17. To be sure, an important school of contemporary media theory has stressed the oppositional moment in popular culture, but this perspective was not yet available for Hill and his staff and is far from universally accepted today.

18. John Downing, *Radical Media,* 76–77.

19. Introduction to "Prospectus for Pacifica Foundation."

20. Ibid.

21. "Pacifica Is Twenty-Five," produced by Larry Josephson, WBAI, Pacifica Foundation, April 1974.

22. Stebbins, *Listener Sponsored Radio,* 61.

23. Ibid.

24. Lewis Hill, "The Theory of Listener-Sponsored Radio," transcribed in McKinney, *The Exacting Ear.*

25. Ibid., 20.

26. Ibid.

27. Ibid.

28. James Baughman, *Republic of Mass Culture,* 19.

29. Ibid., 21 (emphasis added).

30. Hill, "Theory of Listener-Sponsored Radio," 23 (emphasis added).

31. Ibid., 24.

32. Ibid., 24–25.

33. Ibid., 25.

34. Kenneth Rexroth, *An Autobiographical Novel,* 519.

35. McKinney, introduction to *The Exacting Ear,* 10.

36. "The Exacting Ear," WBAI, Pacifica Foundation, 1966.

37. McKinney, introduction to *The Exacting Ear*, 3.

38. "Report to the Executive and Advisory Members of Pacifica Foundation," 35.

39. Ibid., 20.

40. Eleanor McKinney, "The Exacting Ear," WBAI, Pacifica Foundation, 1966.

41. Ibid.

42. Ibid.

43. "Report to the Executive and Advisory Members of Pacifica Foundation," 59.

44. "Is Free Speech Still Free?" produced by Lewis Hill, KPFA, Pacifica Foundation, 1951(?). All the following quotes, except where noted, are from this tape. The tenor of the conversation sometimes makes it difficult to know the person who is speaking, in which case the dialogue is presented but not attributed.

45. Eleanor McKinney, "KPFA History," 8.

46. Ibid., 1 (emphasis in original).

47. Mulford Q. Sibley, *The Obligation to Disobey*, 19.

48. "Pacifica's Birthday."

49. Bill Triest, "Letter to Vera Hopkins, March 21, 1974" (emphasis in the original).

50. Henry Geiger, "The Unfinished Revolution," 1.

51. Ibid., 3.

52. Stebbins, "Listener-Sponsored Radio," 102.

53. See Ralph Engelman, *Public Broadcasting in America*, chap. 3.

54. McKinney, "KPFA History."

55. "Playing in the FM Band," WBAI, Pacifica Foundation, 1974.

56. James Lumpp, "The Pacifica Experience, 1946–1975," 132.

57. Ibid., 134.

58. Ibid., 135.

59. Stebbins, "Listener-Sponsored Radio," 162.

60. Ibid., 163.

61. Ibid.

62. Ibid.

63. Ibid., 184.

64. "Playing in the FM Band," WBAI, Pacifica Foundation, 1974.

65. Lumpp, "The Pacifica Experience," 159.

66. "The Brash Experiment," produced by Lew Hill, KPFA, Pacifica Foundation, 1956.

67. "Progress and Penury."

68. Stebbins, "Listener-Sponsored Radio," 191.

69. Ibid., 194.

70. Ibid., 196.

4. THE DEVELOPMENT OF THE PACIFICA NETWORK

1. "KPFA: A Prospectus of the Pacifica Station."

2. "Pacifica's Forty-Fifth Birthday," produced by Larry Bensky, KPFA, Pacifica Foundation, 15 April 1994.

3. Eugene Stebbins, "Listener-Sponsored Radio," 355.

4. Ibid., 364.

5. "Listener Sponsored Radio in Southern California: A Prospectus," 12.

6. Ibid.

7. Ibid., 24.

8. James Lumpp, "The Pacifica Experience, 1946–1975," chap. 4.

9. "In the Beginning: Schweitzer Memorial Broadcast," produced by Larry Josephson, WBAI, Pacifica Foundation, 1971.

10. Steve Post, *Playing in the FM Band.*

11. "Intellectual's Radio."

12. Ibid.

13. Ibid.

14. *Time,* 24 February 1958.

15. *WBAI Folio,* December 1961.

16. "Birth of Pacifica," WBAI, Pacifica Foundation, 1966.

17. *New York Times,* 10 July 1960.

18. "Purposes and Goals," 3, in *Pacifica Radio Sampler.*

19. Roy Finch, interview by the author, 28 January 1994. Tape in author's possession.

20. "Mass Culture: A Roundtable with Dwight MacDonald, Daniel Bell, and Winston White," KPFA, Pacifica Foundation, 1959. All the following quotes in this section are from this program unless otherwise noted.

21. "KPFK through the Years," produced by Carlos Hagen, KPFK, Pacifica Foundation, 1967.

22. "After the Silent Generation," WBAI, Pacifica Foundation, 1961. All quotes in this section are from this program.

23. *Counterattack* 14, no. 3.

24. Ibid., 16 (emphasis added).

25. Stebbins, "Listener-Sponsored Radio," chap. 2.

26. Robert Schutz, "The Investigation of Pacifica Foundation by the United States Senate Internal Security Subcommittee, 1963."

27. Jerry Shore, interview by the author, 4 March 1994. Notes in author's possession.

28. Steve Post, *Playing in the FM Band,* 22.

29. Ibid.

30. Shore, interview by the author, 4 March 1994.

31. Ibid.

32. Chris Koch, in "Elsa Knight Thompson Remembrance," produced by Chris Koch, KPFA, Pacifica Foundation, 1982.

33. *New Republic,* 2 November 1964.

34. Schutz, "Investigation of Pacifica," 2–3.

35. Ibid., 4–5.

36. Ibid., 5.

37. Ibid., 8.

38. *New York Times,* 15 January 1963.

39. *Communications Act of 1934.*

40. *Borrow v FCC,* 285 F. 2d 666.

41. *New York Times,* 15 November 1963.

42. Schutz, "Investigation of Pacifica," 49.

43. Jerry Shore, interview by the author, 4 March 1994.

44. E. W. Henry, "NAB Speech," KPFA, Pacifica Foundation, 1964.

45. Jerry Shore, interview by the author.

46. Larry Bensky, 13 March 1994, and William Mandell, 4 March 1994, interviews by the author. Notes in author's possession.

47. "Elsa Knight Thompson Remembrance," 1982.

48. Ibid.

49. Ibid.

50. Ibid.

51. "KPFK through the Years," 1967.

52. Quoted in Lumpp, "The Pacifica Experience," 212.

53. Ibid., 210.

54. Ibid.

55. Stebbins, "Listener-Sponsored Radio," 285.

56. William Mandell, interview by the author.

57. Burton White, interview by the author, 9 April 1994. Notes in author's possession.

58. "KPFK through the Years," 1967.

59. "Interview with James Farmer," produced by Elsa Knight Thompson, KPFA, Pacifica Foundation, 1961.

60. "After the Silent Generation."

61. Vera Hopkins, "Unionization at KPFA," 4.

62. Ibid., 5.

63. "Elsa Knight Thompson Remembrance."

64. "*The Turbulent Decade,* Tape 4: Violence in America," WBAI, Pacifica Foundation, 1971.

65. "Saul Alinsky Lecture," WBAI, Pacifica Foundation, 1971.

66. "*The Turbulent Decade,* Tape 1: Militarism and Democracy," WBAI, Pacifica Foundation, 1971.

67. "Robert Hutchins Lecture," KPFA, Pacifica Foundation, 1973.

68. Murray B. Levin, *Talk Radio and the American Dream,* 19.

69. Paul Dallas, *Dallas in Wonderland,* 52–53.

70. Wayne Munson, *All Talk,* 45.

5. Free Speech Radio

1. *Radio Act of 1927.*

2. It is possible to dispute this contention, and certain "laissez-faire" legal scholars and media moguls have claimed excessive, unneeded government interference into broadcasting. That I believe this is a purely ideological argument, based on a tendentious reading of history, need not occupy us.

3. Joy Elmer Morgan, "A National Culture," 28–29.

4. In most law schools today, constitutional law consists of a two-semester sequence: one term on the First Amendment and one on everything else.

5. Vincent Blasi has recently highlighted the twin motifs of Brandeisian First Amendment jurisprudence: the substantive, ancient Greek virtues of "courage" and "participation." See Blasi, "The First Amendment and the Ideal of Civic Courage."

6. *Whitney v California,* 1033.

7. One might well argue that Brandeis's postulates on public discussion not only precede the work on the public sphere initiated by Habermas but anticipate some of the more radical criticism of that category in its insistence on the contentious, revolutionary element of the public discourse.

8. Lewis Hill, *KPFA Folio,* September 1953.

9. Eleanor McKinney, interview by the author, 5 December 1994. Notes in author's possession.

10. Alexander Meiklejohn, *Political Freedom,* 88. Note also how his fury at the way that "freedoms" mask more basic enslavements resembles the form of Marx's famous critique of civil liberties in "On the Jewish Question."

11. Ibid., 26–27.

12. "The First Amendment: Core of Our Constitution," KPFA, Pacifica Foundation, 1957.

13. Meiklejohn, *Political Freedom,* 75 (emphasis added).

14. "Prospectus for the Pacifica Foundation."

15. "Pacifica Radio: Purposes and Goals, August 1946 through September 1977," 3.

16. John Stuart Mill, *On Liberty,* 6.

17. "Hugo Black on the Bill of Rights," KPFA, Pacifica Foundation, 1958.

18. Eleanor McKinney, "The Exacting Ear," WBAI, Pacifica Foundation, 1966.

19. Tom Frank, *The Conquest of Cool.*

20. Norman Mailer, "The White Negro," 313.

21. Walt Whitman, *The Works of Walt Whitman,* 224–25.

22. "The Poetry of Lawrence Ferlinghetti," KPFA, Pacifica Foundation, 1959.

23. Stebbins, "Listener-Sponsored Radio," 159.

24. Ibid.

25. Burton White, interview by the author, 9 April 1994. For White, as for so many others, Elsa Knight Thompson was the pivotal figure in this.

26. "The Berkeley Free Speech Movement," KPFA, Pacifica Foundation, 1966.

27. Ibid.

28. Shana Alexander, "You Don't Shoot Mice with Elephant Guns," 27.

29. Ibid.

30. Quoted in Lumpp, "The Pacifica Experience," 219.

31. Wini Breines, *Community and Organization in the New Left, 1962–1968,* 42.

32. "Obscenity and Pacifica," KPFK, Pacifica Foundation, 1971. All quotes from the hearings referred to in this chapter are from this program and will not be further noted.

33. U.S. Criminal Code: 18 USC 1864.

34. *Roth v United States,* 354 U.S. 476, 77 S. Ct 1304 (1957).

35. George Carlin, *Occupation Foole.*

36. *FCC v Pacifica Foundation,* 438 U.S. 726, 751 (1978).

37. Robert Wolff, "Pacifica's Seven Dirty Words," 971.

38. *FCC v Pacifica Foundation.*

39. Ibid, 3054.

40. James C. Hsiung, "Indecent Broadcast," 55.

41. *FCC v Pacifica Foundation.*

42. Schiffrin, *The First Amendment, Democracy, and Romance,* 204.

43. Ibid.

44. Lewis Hill, "The Theory of Listener-Sponsorship," 22.

45. Joel Feinberg, "Obscene Words and the Law," 141.

46. John Dewey, *Experience and Nature,* 265.

47. Robert Post, "Between Democracy and Community."

6. WBAI AND THE EXPLOSION OF LIVE RADIO

1. "Report to the Executive and Advisory Members of Pacifica Foundation on the Experience of the Radio Station KPFA in Its First Five Months," 3.

2. Vera Hopkins, "Letter to Larry Bensky, August 22, 1983."

3. Ibid. (emphasis added).

4. John Dewey, "The Lost Individual," 385.

5. See, for example, Lorenzo Milam, *The Original Sex and Broadcasting.*

6. Joan Scott, "Multiculturalism and the Politics of Identity," 18.

7. Former WBAI station manager Larry Josephson, interview by the author, 29 July 1993. Notes in possession of the author.

8. Jay Sand, "The Radio Waves Unnameable."

9. *WBAI Folio,* July 1964, 4.

10. Fass, in Jay Sand, "The Radio Waves Unnameable," 6.

11. Wini Breines, *Community and Organization in the New Left, 1962–1968,* 22. Breines was referencing events at Berkeley, but her insights seem well suited to describe Fass's contribution to WBAI as well, a situation not that dissimilar to the circumstances Breines is depicting.

12. Steven Post, *Playing in the FM Band,* 74.

13. Fred Powell, "Switched On Radio: WBAI," 4.

14. Paul McIsaac, interview by the author, 3 September 1994. Tape in the author's possession.

15. Vin Scelsa, in Jay Sand, "The Radio Waves Unnameable," 2.

16. Celeste Wesson, interview by the author, 18 July 1993. Notes in the author's possession.

17. Margot Adler, interview by the author, 22 July 1993. Notes in the author's possession.

18. Arnold Sacher, interview by the author, 3 August 1993. Notes in the author's possession.

19. Former vice president of WBAI Board of Directors Carolyn Goodman, interview by the author, 3 September 1993. Tape recording in author's possession.

20. *WBAI Folio,* July 1972, 2.

21. Josephson, interview by the author.

22. Adler, interview by the author.

23. Josephson, interview by the author.

24. Ibid.

25. Post, *Playing in the FM Band,* 104.

26. Ibid.

27. Ed Goodman, "The State of the Station," 5.

28. Wesson, interview by the author.

29. This discussion at WBAI in 1973 forecasted the direction that PBS has subsequently followed.

30. Ed Woodward, interview by the author, 23 May 1993. Notes in the author's possession.

31. Lisa Ryan, interview by the author, 12 August 1993. Notes in the author's possession.

32. McIsaac, interview by the author. McIsaac considers himself a friend and supporter of Pitts and gay liberation. He does, however, question some of the style of Pitts's programs.

33. See Sand, "Radio Waves Unnameable."

34. Wesson, interview by the author.

35. Ibid.

36. *Village Voice,* 14 February 1977.

37. Wesson, interview by the author.

38. *Village Voice,* 14 February 1977.

39. "Crisis at BAI, Tape 1," WBAI, Pacifica Foundation, 1977.

40. *Village Voice,* 14 February 1977.

41. "Crisis at BAI, Tape 2," WBAI, Pacifica Foundation, 1977.

42. George Fox, interview by the author, 4 September 1994. Notes in the author's possession.

43. Wesson, interview by the author.

44. Bob Fass, interview by the author, 4 August 1994. Notes in the author's possession.

7. BELOVED COMMUNITY

1. Tracy Tyler, ed., *Radio as Cultural Agency.*

2. James Rorty, "Response," 99.

3. It is no coincidence that "public" broadcasting in this country has avoided these pitfalls and opted for safety above democracy ("order at the cost of liberty" in Brandeis's terms), pursuing the strategy first enunciated by educational broadcasters nearly seventy years ago.

4. George Katsiaficas, *The Imagination of the New Left.*

5. David Armstrong, *A Trumpet to Arms: Alternative Media in America,* 79.

6. Engelman, *Public Broacasting in America.*

7. Todd Gitlin, *The Twilight of Common Dreams: Why America Is Wracked by Culture Wars.*

8. Nancy Fraser, *Unruly Practices: Power, Discourse, and Gender in Contemporary Social Theory,* 167–86.

9. Josiah Royce, "The Body and the Members," 230.

10. Fredric Jameson, "Periodizing the Sixties," 178.

11. The most coherent and definitive theoretical statement of this position is Barbara Herrstein Smith's *Contingencies of Value.*

12. Gitlin, *The Twilight of Common Dreams,* 102.

13. David Apter, "Democracy and Emancipatory Movements," 169.

14. Bruce Girard, ed., *A Passion for Radio,* ix.

15. Ibid., 2.

16. Gordon S. Wood, "The Democratization of Mind in the American Revolution," 133–34.

CONCLUSION

1. Ed Goodman, "The State of the Station," 5.

2. "A History of Community Radio, Tape 1: 'It Seemed Important,'" produced by Keith McClure, KPFA, National Association of Community Broadcasters, 1978.

3. William James, "The Moral Equivalent of War."

4. Elaine Scarry, *The Body in Pain.*

5. Brandeis, in *Whitney v. California,* 1034.

6. Ibid., 1033.

7. Scarry, *The Body in Pain,* 90

8. Julius Lester, foreword to Post, *Playing in the FM Band,* xiv.

9. Ibid.

10. Unger, *Passion,* 109.

11. Ibid.

Adler, Margot. *Heretic's Heart: A Journey through Spirit and Revolution.* Boston: Beacon Press, 1997.

Alexander, Shana. "You Don't Shoot Mice with Elephant Guns." *Life,* 15 January 1965.

Appendix to "Report to the Executive and Advisory Members of Pacifica Foundation." In *Pacifica Radio Sampler.*

Apter, David. "Democracy and Emancipatory Movements: Notes for a Theory of Inversionary Discourse." In *Emancipations, Modern and Postmodern,* ed. Jan Nederveen Pieterse. Newbury Park, Calif.: Sage Publications, 1992.

Armstrong, David. *A Trumpet to Arms: Alternative Media in America.* Boston: South End Press, 1981.

"Articles of Incorporation." Pacifica Foundation, 9 August 1946. In *Pacifica Radio Sampler.*

Aylesworth, Martin. "National Broadcasting." In *Radio and Its Future,* ed. Martin Codel. New York: Harper and Brothers, 1930.

Banning, William Peck. *Commercial Broadcasting Pioneer: The WEAF Experiment, 1922–1926.* Cambridge: Harvard University Press, 1946.

Barlow, William. "A History of the Pacifica Radio Network." Unpublished MS, 1993.

Barnouw, Erik. *A Tower in Babel: A History of Broadcasting in the United States to 1933.* New York: Oxford University Press, 1966.

———. *The Golden Web: A History of Broadcasting in the United States.* Vol. 2, *1933–1953.* New York: Oxford University Press, 1968.

Baughman, James. *Republic of Mass Culture.* Baltimore: Johns Hopkins University Press, 1992.

Bender, Thomas. *Community and Social Change in America.* New Brunswick: Rutgers University Press, 1978.

Blasi, Vincent. "The First Amendment and the Ideal of Civic Courage: The Brandeis Opinion in *Whitney v. California*." *William and Mary Law Review* 29, no. 4 (summer 1988).

Bourdieu, Pierre. *Distinction, A Social Critique of the Judgement of Taste.* Trans. Richard Nice. Cambridge: Harvard University Press, 1984.

———. "Social Space and Symbolic Power." *Sociological Theory* 7, no. 1 (spring 1989).

Breines, Wini. *Community and Organization in the New Left, 1962–1968: The Great Refusal.* New Brunswick: Rutgers University Press, 1989.

Brock, Peter. *Twentieth-Century Pacifism.* New York: Van Nostrand Reinhold, 1968.

Carlin, George. *Occupation Foole.* Reprise Records, 1971.

Chaffee, Zechariah. *Free Speech in the United States.* Cambridge: Harvard University Press, 1943.

Chatfield, Charles. *For Peace and Justice: Pacifism in America.* Knoxville: University of Tennessee Press, 1970.

Chickering, Roger. *Imperial Germany and a World without War: The Peace Movement and German Society, 1892–1914.* Princeton: Princeton University Press, 1975.

Counterattack 14, no. 3 (5 February 1960).

Dallas, Paul. *Dallas in Wonderland.* Los Angeles, 1967.

Dellinger, David. Forward to *Against the Tide: Pacifist Resistance in the Second World War, an Oral History.* 1984 WRL Calendar, ed. Deena Hurwitz and Craig Simpson. New York: War Resisters League, 1984.

Dewey, John. *Education and Democracy.* 1916. Reprint, New York: Free Press, 1975.

———. *The Public and Its Problems.* 1927. Reprint, Denver: Allan Swallow, 1957.

———. *Experience and Nature.* 1929. Reprint, New York: Dover, 1965.

———. "The Lost Individual." In *Classical American Philosophy,* ed. John Stuhr. New York: Oxford University Press, 1987.

Douglas, Susan. *Inventing American Broadcasting, 1899–1922.* Baltimore: Johns Hopkins University Press, 1987.

Downing, John. *Radical Media.* Boston: South End Press, 1980.

Eller, Cynthia. *Conscientious Objectors and the Second World War: Moral and Religious Arguments in Support of Pacifism.* New York: Praeger, 1991.

Engelman, Ralph. *Public Broadcasting in America.* New York: Sage, 1996.

EXTRA!: The Magazine of FAIR 7, no. 1 (January–February 1994).

Federal Communications Commission. *Communications Act of 1934.* Washington, D.C.: Federal Communications Commission, 1935.

Feinberg, Joel. "Obscene Words and the Law." *Law and Philosophy* 2 (1983).

Frank, Tom. *The Conquest of Cool.* Chicago: University of Chicago Press, 1997.

Fraser, Nancy. *Unruly Practices: Power, Discourse, and Gender in Contemporary Social Theory.* Minneapolis: University of Minnesota Press, 1989.

Frost, S. E., Jr. *Is American Radio Democratic?* Chicago: University of Chicago Press, 1936.

————, ed. *Education's Own Stations: The History of Broadcast Licenses Issued to Educational Institutions.* Chicago: University of Chicago Press, 1936.

Geiger, Henry. "The Unfinished Revolution." *Manas* 1, no. 1 (7 January 1948). Reprinted in *The Manas Reader.* New York: Grossman Publishers, 1971.

Girard, Bruce, ed. *A Passion for Radio.* Montreal: Black Rose Press, 1992.

Gitlin, Todd. *The Twilight of Common Dreams: Why America Is Wracked by Culture Wars.* New York: Henry Holt, 1995.

Goodman, Ed. "The State of the Station." *WBAI Folio,* January 1973.

Habermas, Jürgen. *The Structural Transformation of the Public Sphere.* Trans. Thomas Burger with the assistance of Frederick Lawrence. Cambridge: MIT Press, 1991.

Harbord, James G. "Radio and Democracy." *Forum* 81 (April 1929).

Herring, E. Pendleton. "Politics and Radio Regulation." *Harvard Business Review* (October 1934).

Hill, Lewis. *Voluntary Listener-Sponsorship.* Berkeley: Pacifica Foundation, 1958.

————. "The Theory of Listener-Sponsored Radio." In *The Exacting Ear,* ed. Eleanor McKinney. New York: Pantheon, 1966.

————. "What Is an Audience?" In "The Pacifica Papers," a special edition of *KPFA Folio,* February 1972.

Hollingdale, R. J., trans. *A Nietzsche Reader.* New York: Penguin, 1982.

Hopkins, Vera. "Letter to Larry Bensky, August 22, 1983." In *Pacifica Radio Sampler.*

————. "Unionization at KPFA." In *Pacifica Radio Sampler.*

————. ed. and comp. *The Pacifica Radio Sampler.* Berkeley: Pacifica Foundation, 1984.

House hearings discussing the proposed Federal Radio Act. *Congressional Record.* 67th Cong., 1st sess., 2 March 1926.

Hsiung, James C. "Indecent Broadcast: An Assessment of Pacifica's Impact." *Communications and the Law.* (1988).

"Intellectual's Radio." *New Yorker,* 19 December 1959.

James, William. "The Moral Equivalent of War." In William James, *Writings, 1902–1910,* ed. Bruce Kuklick. New York: Library of America, 1987.

Jameson, Fredric. "Periodizing the Sixties." In *The Sixties without Apology,* ed. Sohnya Sayres, Anders Stephanson, Stanley Aronowitz, and Fredric Jameson. Minneapolis: University of Minnesota, 1984.

Kateb, George. *The Inner Ocean.* Ithaca: Cornell University Press, 1993.

Katsiaficas, George. *The Imagination of the New Left.* Boston: South End Press, 1987.

Keegan, John. *History of Warfare.* New York: Alfred E. Knopf, 1993.

"KPFA: A Prospectus of the Pacifica Station" (1948). In *Pacifica Radio Sampler.*

KPFA Folio, September 1953. In *Pacifica Radio Sampler.*

Laclau, Ernesto. *Reflections on the Revolutions of Our Time.* New York: Verso, 1992.

Lerner, Max, ed. *The Mind and Faith of Justice Holmes.* New York: Modern Library, 1943.

Lester, Julius. Foreword to S. Post, *Playing in the FM Band.*

Letters. *WBAI Folio,* July 1972.

Levin, Murray B. *Talk Radio and the American Dream.* Lexington, Mass.: Lexington Books, 1987.

"Listener Sponsored Radio in Southern California: A Prospectus." In *Pacifica Radio Sampler.*

Lumpp, James. "The Pacifica Experience, 1946–1975: Alternative Radio in Four United States Metropolitan Communities." Ph.D. diss., University of Missouri–Columbia, 1977.

MacDonald, J. Fred. "Government Propaganda in Commercial Radio: The Case of Treasury Star Parade, 1942–1943." *Journal of Popular Culture* 2 (fall 1979).

Mailer, Norman. "The White Negro." In *Advertisements for Myself.* New York: G. P. Putnam, 1959.

Marvin, Carolyn. *When Old Technologies Were New.* New York: Oxford University Press, 1987.

Marx, Karl. *Capital.* Vol. 1. Trans. Ben Fowkes. New York: Vintage Books, 1977.

McChesney, Robert W. *Telecommunications, Mass Media, and Democracy: The Battle for the Control of U.S. Broadcasting, 1928–1935.* New York: Oxford University Press, 1993.

McKinney, Eleanor. Introduction to *The Exacting Ear,* ed. Eleanor McKinney. New York: Pantheon, 1966.

———. "KPFA History." In *Pacifica Radio Sampler.*

Meiklejohn, Alexander. *Political Freedom: The Constitutional Powers of the People.* 1948. Reprint, New York: Harper and Brothers, 1960.

Milam, Lorenzo. *The Original Sex and Broadcasting: A Handbook on Starting a Radio Station for the Community.* San Diego: MHO and MHO, 1988.

Mill, John Stuart. *On Liberty.* Ed. David Spitz. New York: Norton, 1987.

Morgan Joy Elmer. "A National Culture — By-Product or Objective of National Planning?" In *Radio as a Cultural Agency,* ed. Tracy Tyler. Washington, D.C.: National Committee on Education by Radio, 1934.

Morris, Jimmy. *The Remembered Years: A Personal View of the 50 Years of Broadcasting That We Call KOAC, Corvallis Service to People of Oregon from 1922 through 1972 with Program Philosophies, Historical Evolution of Facilities — and the People Who Kept It All Happening.* Corvallis, Oreg.: A Continuing Education Book, 1972.

Munson, Wayne. *All Talk: The Talkshow in Media Culture.* Philadelphia: Temple University Press, 1993.

NACRE. *Four Years of Network Broadcasting: A Report by the Committee on Civic Education by Radio of the National Advisory Council on Radio in Education and the American Political Science Association.* Chicago: University of Chicago Press, 1937.

National Association of Broadcasters. *Broadcasting in the United States.* Washington, D.C.: National Association of Broadcasters, 1933.

"National Committee on Education by Radio." In *Education by Radio* 1 (25 June 1931): 3.

"NSC-68: A Report to the National Security Council, by the Executive Secretary on United States Objectives and Programs for National Security." In *Containment*, ed. Thomas H. Etzhold and John Lewis Gaddis. New York: Columbia University Press, 1978.

"Pacifica Radio: Purposes and Goals, August 1946 through September 1977." In *Pacifica Radio Sampler.*

"Pacifica's Birthday." *San Francisco Bay Guardian*, 17–31 March 1974.

Post, Robert. "Between Democracy and Community." In *Nomos XXXV: Democratic Community*, ed. John Chapman and Ian Shapiro. New York: New York University Press, 1993.

Post, Steve. *Playing in the FM Band: A Personal Account of Free Radio.* New York: Viking, 1974.

Powell, Fred. "Switched On Radio: WBAI." *Avant-Garde Magazine*, n.d.

"Progress and Penury." *KPFA Folio*, 7 February–2 March 1957.

"Prospectus for the Pacifica Foundation" (1946). In *Pacifica Radio Sampler.*

"Report to the Executive and Advisory Members of Pacifica Foundation on the Experience of the Radio Station KPFA in Its First Five Months." Berkeley, 1949. In *Pacifica Radio Sampler.*

Rexroth, Kenneth. *An Autobiographical Novel.* Rev. ed. New York: Pantheon, 1991.

Rorty, James. *Order on the Air.* New York: John Day Press, 1933.

———. *Our Master's Voice.* New York: John Day Press, 1934.

———. "Response." In *Radio as Cultural Agency*, ed. Tracy Tyler. Washington, D.C.: National Committee on Education by Radio, 1934.

Royce, Josiah. "The Body and the Members." In *Classical American Philosophy*, ed. John Stuhr. New York: Oxford University Press, 1987.

Sand, Jay. "The Radio Waves Unnameable: BAI, Bob Fass, and Listener Sponsored Yippie!" Senior history thesis, University of Pennsylvania, 1993.

Sarnoff, David. *Looking Ahead.* New York: McGraw-Hill, 1968.

Scarry, Elaine. *The Body in Pain: The Making and Unmaking of the World.* New York: Oxford University Press, 1985.

Schiffrin, Steven. *The First Amendment, Democracy, and Romance.* Cambridge: Harvard University Press, 1990.

Schmitt, Karl. *The Concept of the Political.* Translated, with an introduction and notes by George Schwab. New Brunswick, N.J.: Rutgers University Press, 1976.

Schutz, Robert. "The Investigation of Pacifica Foundation by the United States Senate Internal Security Subcommittee, 1963." In *Pacifica Radio Sampler.*

Scott, Joan. "Multiculturalism and the Politics of Identity." *October* 61 (summer 1992).

Severin, Werner. "Commercial vs. Non-commercial Radio during Broadcasting's Early Years." *Journal of Broadcasting* 20 (fall 1978).

Sibley, Mulford Q. *The Obligation to Disobey*. New York: Council on Religious and International Affairs, 1970.

Smith, Barbara Herrnstein. *Contingencies of Value*. Cambridge: Harvard University Press, 1987.

Stebbins, Eugene. "Listener-Sponsored Radio: The Pacifica Stations." Ph.D. diss., Ohio State University, 1968.

Summers, Harrison B., ed. *A Thirty-Year History of Programs Carried on National Radio Networks in the United States, 1926–1956*. New York: Arno Press and the *New York Times*, 1971.

Unger, Roberto. *Passion: An Essay on Personality*. New York: Free Press, 1984.

Tracy, Jim. "Forging Dissent in an Age of Consensus: Radical Pacifism in America, 1940 to 1970." Ph.D. diss., Stanford University, 1993.

Triest, Bill. "Letter to Vera Hopkins, March 2, 1974." In *Pacifica Radio Sampler.*

Tyler, Tracy, ed. *Radio as Cultural Agency*. Washington, D.C.: National Committee on Education by Radio, 1934.

Village Voice, 14 February 1977.

Walker, Samuel. *In Defense of Civil Liberties: A History of the ACLU*. New York: Oxford University Press, 1990.

Whitman, Walt. *The Works of Walt Whitman in Two Volumes as Prepared by Him for the Deathbed Edition*. Vol. 2, *The Collected Prose*. With a prefatory note by Malcolm Cowley. United States: Minerva Press, 1969.

———. "The Real War Will Never Get in the Books." In *Complete Poetry and Collected Prose*, ed. Justin Kaplan. New York: Library of America, 1982.

WBAI Folio, July 1964.

White, Llewellyn. *The American Radio*. Chicago: University of Chicago Press, 1948.

Whitney v California, 274 U.S. 357, 47 S. Ct. 641. In *Constitutional Law,* ed. Gerald Gunther. 12th ed. Westbury, N.Y.: Foundation Press, 1991.

Williams, Raymond. *The Sociology of Culture*. New York: Oxford University Press, 1983.

Wittner, Lawrence. *Rebels against War: The American Peace Movement, 1933–1983*. Philadelphia: Temple University Press, 1984.

Wolff, Robert. "Pacifica's Seven Dirty Words: A Sliding Scale of the First Amendment." *Law Forum* 1979, no. 4 (1971).

Wood, Gordon. *The Creation of the American Republic*. Chapel Hill: University of North Carolina Press, 1969.

———. "The Democratization of Mind in the American Revolution." In *The Moral Foundations of the American Republic*, 3d ed., ed. Robert H. Horwitz. Charlottesville: University Press of Virginia, 1986.

Zetterbaum, Marvin. *Tocqueville and the Problem of Democracy*. Stanford: Stanford University Press, 1976.

INTERVIEWS

All interviews were conducted by the author.

Adler, Margot. 22 July 1993.

Bensky, Larry. 13 March 1994.

Fass, Bob. 23 December 1993.

Finch, Roy. 28 January 1994.

Goodman, Carolyn. 3 September 1993.

McKinney, Eleanor. 5 December 1994.

Josephson, Larry. 29 July 1993.

Mandell, William. 4 March 1994.

McIsaac, Paul. 3 September 1994.

Ryan, Lisa. 12 August 1993.

Post, Steve. 2 September 1993.

Sacher, Arnold. 3 August 1993.

Shore, Jerry. 4 March 1994.

Wesson, Celeste. 18 July 1993.

White, Burton. 9 April 1994.

Woodward, Ed. 23 May 1993.

Note: In many instances, scant information is available about the producers or date and station of broadcast of these tapes.

1951(?)

"Is Free Speech Still Free?" Produced by Lewis Hill, KPFA, Pacifica Foundation.

1956

"The Brash Experiment." Produced by Lewis Hill, KPFA, Pacifica Foundation.

1957

"The First Amendment: Core of Our Constitution." KPFA, Pacifica Foundation.

1958

"Hugo Black on the Bill of Rights." KPFA, Pacifica Foundation.

1959

"Mass Culture: A Roundtable with Dwight MacDonald, Daniel Bell, and Winston White." KPFA, Pacifica Foundation.
"The Poetry of Lawrence Ferlinghetti." KPFA, Pacifica Foundation.

1961

"After the Silent Generation." WBAI, Pacifica Foundation.
"Interview with James Farmer." Produced by Elsa Knight Thompson, KPFA, Pacifica Foundation.

1964

"E. W. Henry, 'NAB Speech.' " KPFA, Pacifica Foundation.

1965

"KPFA's Sixteenth Birthday." KPFA, Pacifica Foundation. 15 April.
"Nonviolence in a Violent World." KPFA, Pacifica Foundation.

1966

"The Berkeley Free Speech Movement." KPFA, Pacifica Foundation.
"Birth of Pacifica." WBAI, Pacifica Foundation.
"The Exacting Ear." Produced by Eleanor McKinney, WBAI, Pacifica Foundation.

1967

"KPFK through the Years." KPFK, Pacifica Foundation.
"KPFK through the Years." Produced by Carlos Hagen, KPFK, Pacifica Foundation.

1971

"In the Beginning: Schweitzer Memorial Broadcast." Produced by Larry Josephson, WBAI, Pacifica Foundation.
"Obscenity and Pacifica." KPFK, Pacifica Foundation.
"Saul Alinsky Lecture." WBAI, Pacifica Foundation.
"*The Turbulent Decade,* Tape 1: Militarism and Democracy." WBAI, Pacifica Foundation.
"*The Turbulent Decade,* Tape 4: Violence in America." WBAI, Pacifica Foundation.

1973

"Robert Hutchins Lecture." KPFA, Pacifica Foundation.

1974

"Pacifica Is Twenty-Five." Produced by Larry Josephson, WBAI, Pacifica Foundation. April.
"Playing in the FM Band." WBAI, Pacifica Foundation.

1977

"Crisis at BAI, Tape 1 and 2." WBAI, Pacifica Foundation.

1978

"*A History of Community Radio,* Tape 2: 'It Seemed Important.' " Produced by Keith McClure, KPFA, National Association of Community Broadcasters.

1982

"Elsa Knight Thompson Remembrance." Produced by Chris Koch, KPFA, Pacifica Foundation.

"Kenneth Rexroth Remembrance." KPFA, Pacifica Foundation.

1994

"Pacifica's Forty-Fifth Birthday." Produced by Larry Bensky, KPFA, Pacifica Foundation. 15 April.

Jeff Land, a longtime activist in grassroots ecological politics, has taught at the secondary and university levels since 1979. He has produced several films, including a documentary about Los Angeles poet Wanda Coleman, *Mad Dog Black Lady.*